About the Author

Kevin Hile is a Michigan-based author and editor who has written books on a wide range of subjects. His authored titles include *Animal Rights, The Trial of Juveniles as Adults, Dams and Levees, Cesar Chavez, Centaurs, Ghost Ships, ESP,* and *Little Zoo by the Red Cedar: The Story of Potter Park Zoo*. As an editor, he has worked with Visible Ink on a variety of science-related books, including *The Handy Math Answer Book, The Handy Geology Answer Book, The Handy Anatomy Answer Book,* and *The Handy Anatomy Answer Book.* He lives in Mason, Michigan.

ALSO FROM VISIBLE INK PRESS

The Handy Anatomy Answer Book
by James Bobick and Naomi Balaban
ISBN: 978-1-57859-190-9

The Handy Answer Book for Kids (and Parents) 2nd edition
by Gina Misiroglu
ISBN: 978-1-57859-219-7

The Handy Astronomy Answer Book
by Charles Liu
ISBN: 978-1-57859-193-0

The Handy Biology Answer Book
by James Bobick, Naomi Balaban, Sandra Bobick, and Laurel Roberts
ISBN: 978-1-57859-150-3

The Handy Dinosaur Answer Book, 2nd edition
by Patricia Barnes-Svarney and Thomas E Svarney
ISBN: 978-1-57859-218-0

The Handy Geography Answer Book, 2nd edition
by Paul A. Tucci and Matthew T. Rosenberg
ISBN: 978-1-57859-215-9

The Handy Geology Answer Book
by Patricia Barnes-Svarney and Thomas E. Svarney
ISBN: 978-1-57859-156-5

The Handy History Answer Book, 2nd edition
by Rebecca Nelson Ferguson
ISBN: 978-1-57859-170-1

The Handy Math Answer Book
by Patricia Barnes-Svarney and Thomas E Svarney
ISBN: 978-1-57859-171-8

The Handy Ocean Answer Book
by Patricia Barnes-Svarney and Thomas E Svarney
ISBN: 978-1-57859-063-6

The Handy Philosophy Answer Book
by Naomi Zack
ISBN: 978-1-57859-226-5

The Handy Physics Answer Book
by P. Erik Gundersen
ISBN: 978-1-57859-058-2

The Handy Politics Answer Book
by Gina Misiroglu
ISBN: 978-1-57859-139-8

The Handy Religion Answer Book
by John Renard
ISBN: 978-1-57859-125-1

The Handy Science Answer Book™ Centennial Edition
by The Science and Technology Department Carnegie Library of Pittsburgh
ISBN: 978-1-57859-140-4

The Handy Supreme Court Answer Book
by David L Hudson, Jr.
ISBN: 978-1-57859-196-1

VISIT US AT WWW.VISIBLEINK.COM

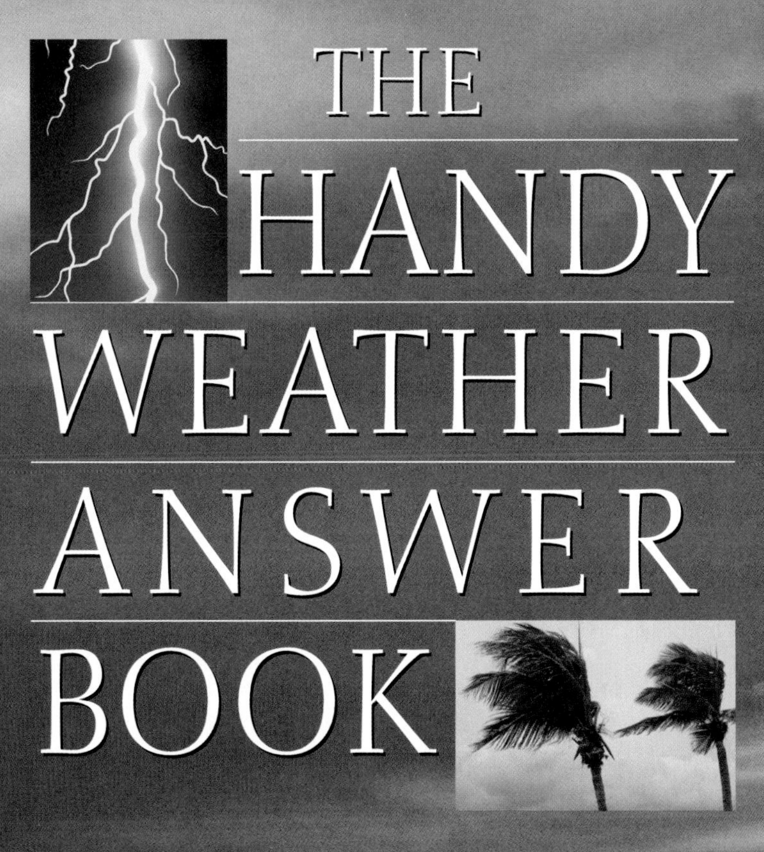

THE HANDY WEATHER ANSWER BOOK

SECOND EDITION

KEVIN HILE

Detroit

THE HANDY WEATHER ANSWER BOOK

Visible Ink Press®
43311 Joy Rd., #414
Canton, MI 48187-2075

Visible Ink Press is a registered trademark of Visible Ink Press LLC.

Most Visible Ink Press books are available at special quantity discounts when purchased in bulk by corporations, organizations, or groups. Customized printings, special imprints, messages, and excerpts can be produced to meet your needs. For more information, contact Special Markets Director, Visible Ink Press, www.visibleink.com, or 734-667-3211.

Managing Editor: Kevin S. Hile
Art Director: Mary Claire Krzewinski
Typesetting: Marco Di Vita
Proofreaders: Amy Marcaccio Keyser

ISBN 978-1-57859-221-0

Cover images: iStock.com

Library of Congress Cataloging-in-Publication Data

Hile, Kevin.
The handy weather answer book / Kevin Hile. – 2nd ed.
p. cm.
Includes bibliographical references and index.
ISBN-13: 978-1-57859-221-0
ISBN-10: 1-57859-221-6
1. Meteorology–Miscellanea. 2. Weather–Miscellanea.
3. Climatology–Miscellanea. I. Title.
QC870H55 2009
551.6–dc22 2009012140

Printed in Singapore

10 9 8 7 6 5 4 3 2 1

Contents

Introduction

It is no exaggeration to say that weather affects everything we do in our lives. Weather influences how we dress, changes our plans for outdoor activities, cancels sporting events, closes airports, changes the course of wars, erodes mountains, destroys entire towns and cities, and has even been blamed for the death of U.S. President William Henry Harrison and the fiery 1986 crash of the Space Shuttle *Challenger*.

While inclement weather might cause us discomfort or even death, our very lives depend on it to sustain agriculture and to keep our bodies healthy. Without weather, the Earth's atmosphere would remain stagnant, rivers and lakes would dry up, and it would be hard to imagine any life thriving on our planet's continents and islands. On the lighter side, weather provides us with a lot of fun: because of weather, we can fly a kite, go skiing, have a snowball fight, or experience the simple joy of splashing in a fresh puddle of rain water.

Because of its power and potential for both harm and good, the weather has been a subject of intense interest and scrutiny by human beings since ancient times. The American humorist Mark Twain once said, "Everybody talks about the weather, but nobody does anything about it." That's not entirely true. People have tried to predict it, even manipulate and change it, for thousands of years, but usually to know great effect. Native American shamans, for example, were known for performing "rain dances" in the hope of causing rain to fall; rain dances have been a cultural part of many other civilizations, too, ranging from ancient Egypt to modern-day life in the Balkans. The ancient Greeks considered weather so important that control of rain and lightning was accredited to Zeus, the king of the gods. The Greeks would therefore pray to Zeus on matters regarding the weather. Of course, with the establishment of the monotheistic religions of Judaism, Christianity, and Islam, control of the weather was regarded as something only God could command.

Philosophers and scientists have long struggled to comprehend the complexities of the weather. Early Greeks, such as Aristotle and Theophrastus of Eresus, mixed in a good deal of conventional wisdom and traditional beliefs with their own efforts to explain and predict weather. With the Renaissance, the Age of Reason, and the Industrial Revolution, science, with the aid of more sophisticated instruments ranging from thermometers and barometers to satellites and Doppler radar, began to measure and analyze the weather more precisely and come up with better theories about cloud formation, temperature, air pressure, and so on.

Despite steadily improving modern technology, predicting the weather is still, in many ways, a haphazard occupation. Some people joke that meteorologists are the only professionals who can keep their jobs and still be wrong half the time. This is really an unfair criticism, though, because modern meteorology has made notable improvements in the critical discipline of predicting severe weather, including hurricanes and tornadoes. Because of efforts by such organizations as the National Weather Service, many lives have been saved in recent decades.

Yet it seems unlikely that we will ever get to the point of being able to predict the weather with 100 percent accuracy. Indeed, according to chaos theory, this is an impossible goal. If, as has been said, a butterfly flapping its wings in China can eventually give birth to a tornado in Oklahoma, what chance do we have of predicting the weather? Because this task seems so hopeless, some people have tried to change the weather directly. For example, scientists have studied cloud seeding with the goal of making it rain in places experiencing prolonged droughts.

Humanity has, indeed, changed the weather. But, as most environmentalists assert, we have done so mostly by accident, and not necessarily for the better. Climate change, ozone holes, and global warming have become catch phrases that inspire great concern among scientists, politicians, and people in general. The pollution of our modern civilization, including carbon monoxide, carbon dioxide, methane, CFCs, and other chemical compounds resulting from industry, agriculture, automobiles, and other sources have been blamed. Many worry that if we don't do something immediately, sea levels will rise, droughts and violent storms will plague humanity, and mass population migrations will result in wars over land, food, and other resources. Still others believe that we are already past the point of no return and climate change is already here today.

Without a firm grasp of meteorology, climatology, hydrology, and other related fields, it is easy to feel ignorant and overwhelmed about what is going on in the current debates on our changing weather. *The Handy Weather Answer Book* is designed to answer your questions in an easy-to-understand format. This book is divided into several chapters by topic, and, all together, it answers over 1,000 questions, ranging from the fundamentals to the cutting-edge of science.

The questions and answers presented here not only cover the usual topics we think about when the subject of weather is brought up (rain, snow, drought, temperature, tornadoes, etc.), but also other phenomena that are related to or affect the weather in some way. For this reason, *The Handy Weather Answer Book* also addresses such areas as atmospheric phenomena, the effects of geographical changes and the oceans on the weather, how our outer space neighborhood influences weather, and theories about climate change.

The Handy Weather Answer Book will take the mystery out of meteorology and, hopefully, inject a bit of fun and excitement into the topic, as well. If you get truly inspired by the subject of weather, the last chapter of this book offers some advice and information about careers in meteorological sciences in case you wish to pursue a formal education in the field.

Many people grouch about the weather. Some even move their places of residence in an effort to avoid it. But a true understanding of the weather can also lend itself to an appreciation of nature and the power behind it. The aesthetic person can discover the beauty of God in a snowflake; the scientist can marvel at the physics behind a twister and the swirl of a hurricane; and all humanity can be humbled by the wild spirit that is weather, the force that refuses to be tamed. As the British author George Robert Gissing once put it:

> For the man sound of body and serene of mind there is no such thing as bad weather; every day has its beauty, and storms which whip the blood do but make it pulse more vigorously.

—Kevin Hile

Acknowledgments

I wish to thank the following people for their assistance in producing this book: Chris Burt for his expertise as a professional meteorologist in reviewing the facts and figures presented in these pages; Larry Baker for his amazing skills as an indexer; Amy Marcaccio Keyser for proofreading the manuscript; Marco Di Vita for typesetting; Mary Claire Krzewinski for her page designs and cover art; and VIP publisher Roger Jänecke for giving me the opportunity to write it all down.

TERMS TO KNOW

What is **weather**?

Weather is defined as the state of the atmosphere at a given location and over a relatively short period of time.

What **factors affect** the **weather**?

It has been said that a butterfly flapping its delicate wings in China will set off a series of events that will eventually result in a hurricane in the Gulf Coast. Weather is extremely complex, so much so that weather forecasting is a highly speculative profession. Some people joke that being a weather broadcaster is the only job you can find where you can be wrong half the time and still stay employed. Weather is affected by temperature, atmospheric composition, land formations, radiation, plate tectonics, geothermic energy, solar winds, biological processes from plants and animals, pollution, and more. All of these factors are considered in this book.

What is **meteorology**?

Meteorology is the scientific study of the weather and, more specifically, how changes in the weather may be forecasted.

What is **hydrology**?

Hydrology is the scientific study of Earth's water supplies, how they are distributed, and how they move and change. Hydrologists are people concerned with water resources, and their work has applications ranging from civil engineering and city planning to environmentalism and conservation.

What, then, is **hydrometeorology**?

Hydrometeorology is an eight-syllable word meaning the study of the exchange of water between the lower atmosphere and the land below it.

What is **climatology**?

Climatology is the study of the world's climates and how they are changing over time.

What is **bioclimatology**?

Bioclimatology is the study of the effects of the climate on living things. Weather and atmospheric conditions can affect humans in many ways, both positive and negative. Our climate affects our moods, the chemical content of our bodies, the chances of getting a disease, and more. In Europe, awareness of the importance of bioclimatology has resulted in weather forecasts that include warnings of possible health hazards. The United States has not kept up with this pace, but meteorologists in America do often warn of such hazards as air pollution and allergen levels, as well as temperature extremes that might be dangerous because of frostbite or heat stroke.

What is **atmospheric chemistry**?

As the name implies, this is the discipline dealing with how gases and other chemicals and particulates in the atmosphere interact with each other, such as with the formation and destruction of ozone, both in the upper atmosphere and as a ground-dwelling pollutant. Atmospheric chemistry is a very complex science, as the composition of the atmosphere is in constant flux. Content is constantly being introduced from the ground; winds continually shift and flow; and radiation from space interacts with the atmosphere as well. Meteorologists specializing in this field have to understand geology, biology, and industrial pollutants (literally, millions of different industrial chemicals entering the atmosphere daily), among other chemical processes. There is considerable work to be done in atmospheric chemistry, as much of what happens in the atmosphere at a chemical level is little understood.

What is **atmospheric physics**?

A complementary field of study to atmospheric chemistry is atmospheric physics. This discipline has to do with such issues as wave and particle physics, acoustics, spectroscopy, optics, and more. A strong command of mathematics is needed for anyone wishing to specialize in this field. The theoretical work involved has applications in satellite, radar, lidar, and other technologies.

What is **diffraction**?

Diffraction is the phenomenon of how light bends around small objects or through small openings. These objects and openings have to be small enough to interfere with wavelengths of light, and so wavelengths in the red spectrum (longer wavelengths) are more affected by light in the bluer spectrum. Diffraction can cause a blurring of light, as well as causing interference in the transmission of invisible energies, such as radio waves and X-rays.

What is **refraction**?

Refraction refers to how light is bent as it passes from one transparent medium to another (for example, from air to water). This happens because light travels at different speeds, depending on the medium. Refraction is the reason why we see rainbows.

Most people think of spray cans when they talk about aerosols, but to a meteorologist any liquid or solid particle suspended in air is considered an aerosol.

What is an **aerosol**?

Many people, when they hear the word "aerosol," think of a chemical aerosol spray from a can of air freshener or hair spray. The word actually applies to any solid or liquid particles suspended in air. Because they are so small, aerosols tend to float (e.g., clouds), though like everything else they are subjected to gravity, falling at a rate of about four inches (10 centimeters) every 24 hours, unless washed away more quickly by rain.

What are **evaporation** and **transpiration**?

Evaporation, as many people know, is what happens when liquid water changes to a gaseous state, escaping into the surrounding atmosphere. The rate of evaporation can be measured using an evaporimeter. Transpiration refers to the release of water vapor from plants, but can also refer to perspiration and sweat being lost from humans and animals.

What is **convection**?

Convection is the transfer of heat vertically through the atmosphere via a liquid medium (e.g. water droplets).

What is **convergence**?

Convergence occurs when air masses approach each other from different directions. As the masses collide, air pressure between them goes up, which causes air to flow upwards.

What is **inversion**?

When the air temperature rises with altitude rather than cooling, the condition is called an inversion.

What is an **ion**?

Ions are atoms or molecules with a positive or negative charge due to differences in the number of protons (positively charged particles in an atom's nucleus) and electrons (negatively charges particles that "orbit" the nucleus). Meteorologists are interested in ions, especially with regard to the ionosphere, because they are highly reactive with other elements and chemicals in the atmosphere.

What is **plasma**?

Plasma is the fourth state of matter (the other states being solid, liquid, and gas). It is formed when electrons are stripped away from atoms and a mix of free electrons and the resulting ions exist together. Plasma is found in stars, which makes it actually the most common state of matter in the universe. But plasma is also found in the solar winds that blow out from the Sun and collide with the magnetosphere. Some plasma radiation makes its way into the ionosphere, too. Lightning is also a form of plasma.

What is the **azimuth**?

Used in navigation and in reporting the position of stars, planets, and other celestial bodies, the azimuth is the number of degrees between the direction of North (0°) and the direction in which the object is viewed from the perspective of the observer. In more mathematical terms, it is the angle between two vertical planes, one formed between the observer and the object observed, and the other formed by the observer and true North.

ORGANIZATIONS

What is the purpose of the **National Oceanic and Atmospheric Administration** (NOAA)?

NOAA is an agency within the U.S. Department of Commerce that is responsible for monitoring conditions on land and in the seas that have an effect on our weather, climate, and environment. NOAA is, of course, heavily involved in atmospheric research and weather forecasting, but the agency also supports the responsible

The Lansing, Michigan, office of the U.S. Weather Bureau—shown in this circa 1900 photo—was once located at Michigan Agricultural College (now Michigan State University). The Weather Bureau was the forerunner of the National Weather Service. *(NOAA)*

management of fisheries, is concerned with marine commerce, and is involved in studies to prevent coastal erosion, among many other projects. In essence, NOAA is interested in fostering the economic and environmental health of the country, as well as the safety of its citizens, through scientific management of oceanic, coastal, and mainland resources.

What is the **National Weather Service** (NWS)?

Part of NOAA, the NWS was founded in 1870 as the National Weather Bureau; it was renamed the U.S. Weather Bureau in 1891, and became the National Weather Service in 1967. It focuses on providing the citizens of the United States with warnings about possibly dangerous storms and other weather events. The NWS has forecasting centers in 122 locations around the country, including U.S. territories like Guam, American Samoa, and Puerto Rico.

What is the **National Weather Center** (NWC)?

The NWC is a partnership between the National Oceanic and Atmospheric Administration, state organizations, and the University of Oklahoma. It is a scientific endeavor to better understand the weather, especially on a macroscale involving long periods of time and all levels of the atmosphere.

What is the AMS Seal of Approval Program?

The AMS Seal of Approval is given to forecasters in the media who provide useful and accurate information about the weather. Part of the intention here is to recognize broadcast meteorologists who do more than just read National Weather Service copy on the air. The seal, therefore, is a service to audiences so that they may discern whether they are receiving their information from a certified professional, or simply from a news reader. A meteorologist may receive a seal either in radio or television broadcasting. They are eligible for the seal based on the quality of the information they provide, their professionalism, their demonstrated effort to continue their education in the field, and their participation as an AMS member. Their qualifications are reviewed by a certifying board committee. Finally, the seal is not bestowed permanently, but must be renewed annually.

What is the **National Center for Atmospheric Research** (NCAR)?

Established by the National Academy of Sciences in 1956, the NCAR is based in Boulder, Colorado, and is staffed by (mostly) university scientists who use such tools as radar, airplanes, and supercomputers to help the scientific community better understand the many processes that affect weather. The goal is to increase cooperation between universities and draw on their combined resources in order to accomplish what a single university could not do on its own.

What are the **National Centers for Environmental Prediction** (NCEP)?

Part of the National Weather Service, the National Centers for Environmental Prediction include the following centers:

- The *Aviation Weather Center* for monitoring weather conditions that could prove hazardous to airplane and space flights.
- The *Climate Prediction Center* is focused on how climate affects the country, as well as on short-term climate changes.
- The *Environmental Monitoring Center* is a research center studying ways to improve weather-related sciences, including climatology, hydrology, and ocean weather prediction.
- The *Hydrometeorological Prediction Center* provides rain forecasts for the upcoming week.
- The *Ocean Prediction Center* is responsible for issuing ocean weather warnings in the Atlantic and Pacific Oceans north of the 30th degree parallel.
- The *Space Weather Prediction Center* warns of weather conditions on Earth and in space that could put space missions at risk.

The National Severe Storms Laboratory (NSSL) research facility in Norman, Oklahoma, is shown in this circa 1970 photo. *(NOAA Photo Library, NOAA Central Library; OAR/ERL/National Severe Storms Laboratory)*

- The *Storm Prediction Center* keeps a watchful eye on tornadoes, hurricanes, and other hazardous weather within the lower 48 U.S. states.
- The *Tropical Prediction Center* monitors tropical weather systems within the United States, as well as surrounding regions.

What does the **American Meteorological Society** (AMS) do?

The AMS is an organization of professionals, as well as amateurs, in the field of meteorology and atmospheric and oceanic sciences that is intended to foster communication, promote education, and share resources. Those without formal degrees in the field can still be members with the rank of Associate, and a Student membership level is also available to those still in school. The society, headquartered in Boston, Massachusetts, publishes periodicals and books, awards accomplishments in the field, and sponsors conferences and the Seal of Approval Program.

What is the **National Severe Storms Laboratory** (NSSL)?

The NSSL is NOAA's premier research laboratory. Located in Norman, Oklahoma, the NSSL is dedicated to researching and improving weather radar systems, severe weather forecasting, and the science of hydrometeorology.

What is the **World Meteorological Organization** (WMO)?

Because the weather is a matter of international concern affecting all the world's countries, the WMO is a highly valuable organization that promotes the sharing of meteorological data between nations. Formerly the International Meteorological Organization (est. 1873), the WMO was created in 1950; the next year, it came under the aegis of the United Nations. The WMO is interested in severe weather forecasting and in the impacts of human activities on the environment that affect the climate and weather.

What is the **Space Weather Prediction Center** (SWPC)?

Part of the National Weather Service, the SWPC monitors solar and geophysical events that can affect communications, power grids, artificial satellites, and navigational systems.

Is the **National Aeronautics and Space Administration** (NASA) involved in **weather forecasting**?

Since NASA is involved in implementing weather satellites, it obviously is very much involved in weather forecasting. NASA doesn't only concern itself with sending out manned and unmanned missions into the solar system and beyond; it also spends a lot of time observing the Earth. Weather and Earth science satellites gather information about changes in the climate, land use, and in our oceans.

MEASUREMENTS

What is a **triple point**?

The triple point is the temperature at which a substance can exist in equilibrium in all three of its states: gas, liquid, and solid. For pure water, the triple point—at an air pressure of 4.58 millimeters of mercury—is 32.018°F (0.01°C). The term "triple point," however, can also refer to the spot where an occluded front meets a warm front.

Why are there so many **discrepancies** in the **world records of weather**?

The discrepancies in the data reflect the length of time that we use to measure weather phenomena. Some records were set by observing the weather over decades; others only occurred during the span of a few years or months, or even hours or minutes. Discrepancies also exist because of the various types of instruments that were used over the years, and how they were exposed to the elements.

What is **Universal Coordinated Time**?

Meteorologists, as well as many other scientists, use the standard of Universal Coordinated Time (UTC) as a time reference to coordinate their measurements. Also known as Greenwich Mean Time (GMT), because Greenwich, England, is the place where the standard time is set, as well as Zulu—or "Z"—time, UTC employs the 24-hour clock also used by the military. Thus, 0000 UTC indicates midnight and 1200

UTC is noon. A standard set by meteorologists is to make observations every six hours—at 0000, 0600, 1200, and 1800 UTC.

What is an **isobar**?

An isobar is a line indicating on a weather map the point where the air pressure is the same (i.e., lines of equal pressure). Isobars are a convenient way to locate cold and warm fronts on a map and regions of high and low pressure.

What are some other **terms using the prefix "iso-"** that meteorologists use?

"Iso-" is a handy prefix that means "the same" or "equal" (from the Greek "isos"). The following terms all take advantage of this Greek route.

Iso-Term	Meaning
Isobar	Equal change in air pressure
Isobathytherm	Equal depth in water having the same temperature
Isobront	Equal amount of thunder
Isoceraunic	Equal number of thunder storm events
Isochasm	Equal frequency of observing the Aurora Borealis
Isochrone	Equal time for the same occurrence of an event
Isodrosotherm	Equal dew point temperature
Isogon	Equal wind direction
Isohel	Equal sunshine
Isohume	Equal humidity
Isohyet	Equal rainfall
Isokeraun	Equally intense thunderstorms
Isometrics	Equal lines of elevation
Isoneph	Equal cloud cover
Isonif	Equal snowfall
Isopectic	Places where winter frosts and ice form at the same time of year
Isopleth	Any equal line of something
Isopycnal	Equal air density
Isoryme	Equal incidence of frost
Isotach	Equal wind speed
Isothere	Places where average summer temperatures are the same
Isotherm	Equal air temperature

What is an **ombrometer**?

An ombrometer, also called a micropluviometer, is just a technical word meaning a rain gauge.

A C-130 airplane is shown at the South Pole Station in 1978. This type of plane has often been used for NOAA research. (*photo by Commander John Bortniak, courtesy NOAA Corps*)

How is **snowfall measured**?

Snowfall is measured in a very practical and low-tech way: with a ruler. To get a good average indication of snowfall in a selected area, the National Weather Service takes measurements from several locations, instead of just one, and then averages them out. In places where there is often heavy amounts of snow, tall poles are erected that can measure the white stuff when it accumulates up to several feet, or even meters, deep. Snowfall can also be measured using a heated rain gauge, which melts the snow into water, then converts it back to estimated snow levels by using the formula that one inch of rain water roughly equals 10 inches of snow. However, in North America, this method is not really used because it is not very accurate.

There is also the snow pillow method for measuring snow, which uses a scale to measure the weight of the snowfall. In snowier climates, a tool called a snow board (which is not the same as the snowboards used for wintertime fun) is used. A snow board is a two-foot wide by two-foot high piece of plywood that is painted white and put in a location where snow is not likely to drift. The purpose of the white paint is to minimize the melting effects of solar radiation. Snow depth measurements are then taken with a ruler every six hours. The six hour rule is hard and fast. This was made clear in 1997, when an observer for the National Weather Service recorded a snowfall of 77 inches (196 centimeters) in Montague, New York, within a 24-hour period. This would have been a world record, but it was disallowed when investigators learned that the observer recorded measurements every four hours instead of every six.

What is an **acre-foot**?

One acre-foot is equal to 43,560 gallons (164,875 liters) of water, which is what it would take to bury an acre of land in a foot of water. The term is usually used to measure rainfall runoff, reservoir capacity, and irrigation.

How is **sea water salinity** measured?

The amount of salts in sea water is important because it affects ocean currents, which, in turn, affect the world's climate. Sea water contains a variety of dissolved elements, including chlorine, sodium, magnesium, calcium, sulfur, and potassium. In the past, measurements of salinity were taken simply by going out onto the ocean, filling a bucket with sea water, and testing the salt levels by measuring electrical conductivity (the more salts, the quicker electricity flows through the water because there are more ions present). There are also techniques to measure chlorine or other dissolved elements.

More recently, sophisticated equipment has become available for measuring ocean salts remotely. Low-frequency radiometers mounted on C-130 aircraft can scan the ocean during flights, covering over 38 square miles (100 square kilometers) every hour. The European Space Agency plans to launch its Soil Moisture and Ocean Salinity (SMOS) satellite in 2009 to take readings from space using a two-dimensional interferometric radiometer, a new technology that captures images based on microwave radiation emitted at a frequency of 1.4 gigahertz (GHz).

How is **wind speed measured**?

Wind speed is measured with a device called an anemometer, which was an invention of English physicist Robert Hooke (1635–1703). The most commonly used type is the rotating cup anemometer, which uses three or four small cups that spin around a central pole. Modern anemometers of this sort work using electricity and magnets. As the cups spin, a reed switch within the central pole detects each time a magnet in a cup swings by. This sends out an electronic pulse that has been calibrated to calculate wind speed. The data is then transmitted to a weather station.

An early anemometer designed by John Thomas Romney Robinson in 1846. *(photo by Sean Linehan, NOS, NGS, courtesy NOAA)*

What are some **other types of anemometers**?

Besides the rotating cup anemometer, there is the sonic anemometer, swinging-plate (or pressure-plate) anemometer, pressure-tube anemometer, bridled (or windmill) anemometer, and the aerovane. Weather stations often use sonic anemometers, which calculate both wind speed and direction. Four ultrasound transducers are set up in a circle, evenly spaced apart, in two pairs placed across from each other. A transducer will send out an ultrasonic signal to the one directly across from it. Winds blowing across this path will cause the signal to travel faster, slower, or change direction, thus indicating wind conditions. Pressure-plate and pressure-tube anemometers work by the fact that wind blowing against a plate or through a tube will exert a measurable pressure. Aerovanes and windmill anemometers can measure both speed and direction. As the blades on these devices spin, it is possible to calculate wind speed, and both will turn into the oncoming wind, which indicates direction.

How is **wind direction** usually measured?

A wind vane is the common instrument used to discover wind direction. Wind vanes look like windmills mounted on a pole that allows them to rotate toward the direction of the oncoming wind. Historically, wind vanes have often come in decorative models, often with a rooster or some other farm animal mounted on the top. Of course, there are many other ways to discover wind direction, ranging from the primitive (analyzing the direction smoke is blowing or how balloons are moving) to the more sophisticated, such as Doppler sodar (sound radar) and lidar (light radar). Gyroscopes and GPS devices mounted in airplanes can calculate air speed by comparing the indicated speed to the actual distance covered (i.e., the amount of thrust from the airplane's jets or propellers may be slowed or sped up, depending on whether winds are blowing with or against the plane).

What **standard unit of measurement** is used to indicate **wind speed**?

In most forecasts in the United States, wind speed is described in miles per hour. (Outside this country, it would be expressed in kilometers per hour; most other scientists also prefer to use the metric system). However, the Federal Aviation Administration, National Weather Service, and other groups that work with air and ocean travel, will use knots (one knot equals 1.15 miles per hour, or 1.85 kilometers per hour). Internationally, wind is also commonly measured in meters per hour. For vertical wind speeds, meteorologists use microbars per second, which indicates pressure change with altitude over time, or centimeters per second.

EARLY WEATHER HISTORY

What did the **Greeks once speculate** about the **air**?

The Greek philosopher Anaximander (610–546 B.C.E.) speculated—correctly—that air wasn't just nothing, but, in fact, was made of something. However, he went on to suggest that all matter came from air, which could be changed into different states of matter. This idea actually has some basis in truth, since, for example, water can be precipitated out of humid air, and water can evaporate into air. Anaximander just got a little too carried away and took this idea to extremes by saying air could also become fire and a lot of other things.

Who wrote the ***Meteorologica***?

The Greek philosopher Aristotle (384–322 B.C.E.) released his *Meteorologica* around 340 B.C.E. It was this work that gave us the term "meteorology"; in Aristotle's time, the word *meteor* referred not just to extraterrestrial rocks entering the atmosphere but rather to anything up in the sky, including clouds, rain, snow, etc. *Meteorologica* is the first comprehensive text written on the subject, at least in the Western world. Many of the theories expressed in Aristotle's work, however, are based on mythology and other misplaced notions of what causes weather. For instance, the philosopher believed that hurricanes resulted from a "moral conflict" between "evil" and "good" winds.

What was the most **important weather book** to follow *Meteorologica*?

Aristotle's student Theophrastus of Eresus (c. 372–287 B.C.E.) continued his mentor's study of weather with his *On Weather Signs*, a book that became the last word on weather. It was consulted all the way through about the twelfth century, when it was still used by scholars of the Byzantine Empire. As a predictor of weather, the book strove to describe how to tell when rain, wind, and storms were coming. Theophrastus's version of meteorology, though, was still a mix of well-reasoned observation and superstition.

Who first correctly wrote about the **structure of snowflakes**?

This honor goes to Han Ying, a Han Dynasty scholar who published *Moral Discourses Illustrating the Han Text of the Book of Songs* in 135 B.C.E. Han correctly described how snowflakes always take on a hexagonal form of some kind (unless the flakes are broken), even though this six-sided fundamental structure has incredible variety. The Western scientific world would not get this right until the seventeenth century, when German mathematician and astronomer Johannes Kepler (1571–1630) published *A New Year's Gift; or, On the Six-Cornered Snowflake* in 1611. English mathematician and astronomer Thomas Harriot (c. 1560–1621) actually correctly described snowflakes' hexagonal form in 1591, but this description was not made public.

What makes the ***Historia Naturalis*** important in the history of meteorology?

The *Historia Naturalis* was written by Pliny the Elder (23–79 C.E.) and contained, among other scientific observations, an ambitious survey of weather conditions from Rome, Greece, Egypt, and Babylon. As with the earlier *Meteorologica* and *On Weather Signs,* though, it was still an inaccurate mix of objective science and myth-inspired superstition.

Why was **Hero of Alexandria** an important figure in the **history of meteorology**?

We have Hero (c. 10–70 C.E.; also spelled as Heron) to thank for being the first to scientifically prove that air consists of matter. A genius who invented an early steam engine and showed you could harness wind's power with a windmill, Hero showed that air had volume (therefore, matter) with such creations as the pump and the syringe.

What ancient **Chinese book** first discussed the idea of **solar winds**?

Although the Chinese discussed the idea of energy from the Sun in terms of the notion of *qi* energy, the *Book of Jin* observed back in 635 C.E. that comet tails always pointed away from the Sun. The unknown author understood that this was the result not of wind in our atmosphere, but rather from energy emitted by the Sun itself.

Which **Chinese scholar** first hypothesized about **climate change**?

In the eleventh century C.E. Chinese writer Shen Kuo (1031–1095) noticed that bamboo plants were buried in the ground near Shanbei. This region was far too northerly for bamboo to grow in Shen's time, and he therefore reasoned that the climate there had once been very different.

How was **Abu 'Ali al-Hasan ibn al-Haytham** important to meteorology?

Abu 'Ali al-Hasan ibn (965–c. 1039) was a brilliant scientist in many areas, including engineering, physics, philosophy, mathematics, astronomy, anatomy, medicine, philosophy, psychology, and more. He has been called the "Father of Modern Optics" and the "Founder of Experimental Physics," attesting to his many accomplishments. His seven-volume *Book of Optics* (1011–1021) explained principles with applications ranging from ophthalmology to astronomy to meteorology. As it pertains to meteorology, his work is important for explaining such concepts as reflection, refraction, transparency, translucence, radiancy, and optical illusions (e.g., mirages). He made contributions to the study of rainbows and atmospheric density.

What is a **thermoscope** and who **invented it**?

The history of the thermometer goes back to the ancient Greeks. It is not known exactly who invented a working thermometer, but the earliest record has Philo of Byzantium creating what was called a "thermoscope" back in the second century B.C.E. Similarly crude devices using the expansion of water due to temperature were used throughout the centuries. The prolific Renaissance inventor and artist Galileo Galilei (1564–1642) improved the air thermoscope in 1593. The thermoscope he created uses a different approach to measuring temperatures than the thermometer. Instead of containing a fluid, such as mercury, that is sensitive to changes in heat and cold, the thermoscope suspends several objects within a transparent tube.

Did the ancient Mayans study the weather?

Many people are familiar with the Mayans' interest in calendars and astronomy, but they were also fascinated by the weather. Sometime between 1200 and 1400 C.E. they constructed a lighthouse in what is now Cozumel, Mexico, called the "Tumba del Caracol." The Mayans put candles in the lighthouse, which served the traditional function of warning ships that they were close to land. In addition, at the top of this lighthouse, the clever Mayans strategically placed a variety of seashells. Depending on wind speed and direction, the shells would whistle at different pitches. Depending on which shells were whistling and at what pitch—and their knowledge of what conditions produced storms—the Mayans are said to have been able to predict storms approaching from the Caribbean.

The objects are small glass spheres containing various amounts of liquid and gas and also attached to a piece of metal that is suspended from each one. These floats have varying levels of buoyancy, which could be finely adjusted further by changing the size of the piece of metal attached. Galileo understood that water's density changed with temperature, and so the buoys (distinguished by the color of the dyed fluid inside them) would rise or fall within the tube accordingly. You could tell the temperature based on which buoys were floating and which ones had sunk to the bottom of the tube. In 1610, Galileo replaced the water in the tube with wine (alcohol). Galileo's friend Santorio Santorio (1561–1636) adapted the thermometer to measure body temperature in his medical practice).

Who invented the **modern thermometer**?

Ferdinand II de Medici (1610–1670), Grand Duke of Tuscany, was also an accomplished physicist. He is generally credited with inventing the first modern thermometer in 1641. It consisted of a sealed tube containing alcohol. This type of thermometer was called a "spirit" thermometer, possibly because alcoholic drinks are sometimes referred to as spirits. Today, alcohol thermometers are still referred to by this quaint label. Ferdinand II improved on his design in 1654; ten years later, Robert Hooke (1635–1703) adapted the duke's thermometer, standardizing the measurements in a more logical way (the duke had arbitrarily divided his thermometer into 50 degrees), using the freezing and boiling points of water as standards.

What did **Sir Isaac Newton** contribute to the science of meteorology?

In terms of meteorology, Sir Isaac Newton (1642–1727) is mostly a significant figure because the science of physics and his laws of motion are essential to an understanding of how weather works. Not many people know, though, that Newton also was into rainbows. He was the first to demonstrate how white light is broken up into its spectrum when it passes through a glass prism.

Benjamin Franklin is famous in weather lore for his experiments with lightning, but he also made other contributions to meteorology. (*NOAA*)

What did **Benjamin Franklin** contribute to the science of **meteorology**?

Benjamin Franklin (1706–1790), who is said to have discovered electricity by flying a kite in a storm and who later invented the lightning rod, made the important discovery that low pressure systems caused the atmosphere to circulate in a rotating pattern. He made this discovery in 1743, after unsuccessfully attempting to view an eclipse on October 21. There was a storm in Philadelphia at the time, but he later learned that the skies were clear in Boston that day. Of course, he wasn't able to take an airplane to Massachusetts, but what he did find out the next day was that the storm that had been in Philadelphia had traveled to Boston. From this information, he surmised that the storm was traveling in a clockwise manner from southwest to northeast. Putting two and two together, Franklin concluded that the low pressure system was causing the storm to move in this manner.

Which of America's **founding fathers** were fascinated by meteorology?

Among his many other interests, ranging from agriculture to architecture, law, and politics, Thomas Jefferson (1743–1826) was also fascinated by the weather. Jefferson was offended by the French naturalist Georges Louis Leclerc de Buffon's (1707–1788) assertion that American's were negatively impacted by their climate, making them somehow inferior to Europeans. To prove him wrong, Jefferson and his friend and fellow Founding Father, James Madison (1751–1836), decided to study the weather in earnest. Jefferson made daily observations from his Virginia home at Monticello from 1772 to 1778, and Madison followed his lead from 1784 to 1802. While it might seem painfully obvious today, it was Madison who broke with English logic that said temperature readings should be done indoors; he took the unheard of step of placing his thermometer outside. Today, universities are using Madison's measurements of temperature and precipitation for comparative studies on climate change.

Who was named the United States's **first official meteorologist**?

James P. Espy (1785–1860) was most noted as the author of *The Philosophy of Storms* (1841). A year after this book's publication, the U.S. Congress named him the federal government's meteorologist. He is credited with giving the first accurate description of how thermodynamics plays a role in cloud formation, also explaining the dynamics of low-pressure systems.

Pioneering meteorologist James Espy made discoveries in how thermodynamics influences cloud formation. (*NOAA*)

What was the **Great Exhibition of 1851**?

Held from May 1 through 15, 1851, in London, England, the Great Exhibition of the Works of Industry of All Nations was the first of what would become the World's Fair international exhibits. It was also called the Crystal Palace exhibit, because of the building in which it was held in Hyde Park. Among the many exhibits, the first weather map was displayed there, as well as the "Tempest Prognosticator," a leech barometer invented by George Merryweather.

What was the first **organized network** of **meteorological observatories**?

In 1855 Urbain Jean Joseph Leverrier (1811–1877), a French astronomer, organized an effort to establish weather observatories throughout Europe that would share meteorological data in the first cooperative system of its kind. In 1863 telegraphs linked many of these weather stations together through a central hub in France.

Who was **Cleveland Abbe**?

Also famous as the person who proposed the creation of time zones, Cleveland Abbe (1838–1916) was an American meteorologist and founder of the *Weather Bulletin* (est. 1869), the first daily periodical to include weather forecasts. He also established the National Weather Bureau in 1870, which is now the National Weather Service.

What **newspaper** was the first to begin publishing **daily weather forecasts**?

The London *Times* was the first newspaper to publish daily weather forecasts in 1860. The forecasts were originally written by retired Admiral Robert FitzRoy (1805–1865), who at the time was head of the meteorological department at the Board of Trade. The early reports concerned temperature, air pressure, and rainfall; beginning in 1861, storm forecasts were added.

Cleveland Abbe founded the *Weather Bulletin* and established the National Weather Bureau. (*NOAA*)

Who first discussed the link between **climate change** and how **gases in the atmosphere** absorb heat?

In 1884, American physicist and astronomer S.P. Langley (1834–1906) was the first to publish a scientific paper on how gases in the atmosphere can absorb heat, which has an effect on the Earth's climate.

Who was **Alexander Buchan**?

The most prominent meteorologist of the nineteenth century, Scottish scientist Alexander Buchan (1829–1907) is sometimes referred to as the "Father of Meteorology." He is credited with making great advances in weather charts, including his use of isobars to connect areas of equal pressure in lines that are now familiar to anyone who has seen a weather map; he also understood the importance of ocean and atmospheric circulation like no one else of his age. In his 1868 book, *Handy Book of Meteorology,* he made long-range weather predictions, the first person to do so in a printed publication. Among his most famous ideas was what are now called "Buchan Spells." These are predictable blips—abrupt changes in temperature—in the usually smooth transition in weather between the seasons. For instance, he predicted that a cold Buchan Spell typically occurred the week before Valentine's Day. Buchan was wise enough, though, to know that such a rule could never be hard and fast, and admitted that his Buchan Spells allowed for some variations and sometimes never occur at all.

THE SEASONS

When do the **seasons start and end**?

When it comes to climate and weather, the seasons start at different times of year depending on where one is on Earth. Astronomically speaking, though, the first day of spring happens on the vernal equinox; the first day of summer happens on the summer solstice; the first day of fall happens on the autumnal equinox; and the first day of winter happens on the winter solstice.

When it comes to official weather statistics, the seasons are considered to be as follows: winter is December through February; spring is March through May; sum-

How does weather affect Earth's rotation?

Imagine the water, clouds, and other gases lying on top of the Earth's crust as a big soupy mass that can shift around as the planet rotates on its axis and is tugged on by the Moon, Sun, and other planets. The oceans and atmosphere slosh around due to tidal action, bulging a bit on one side or the other, and this can hamper or speed up the planet's motion. In comparison to the total weight of our planet, the liquids and gases are fairly light, but the inertia they experience does, in fact, change Earth's speed. The amount of change is not noticeable to us: a few thousandths of a second each year. Over millions of years, however, this has a cumulative effect.

mer is June through August; and fall is September through November. So, if you hear a report, for example, that "last summer was the hottest on record," that means June 1 through August 31, and not June 21 through September 21, which is how it is marked on your typical calendar.

What is the **ecliptic plane**?

The ecliptic plane is the plane of Earth's orbit around the Sun. Ancient astronomers were able to trace the ecliptic as a line across the sky, even though they did not know Earth actually orbited the Sun. They merely followed the position of the Sun compared to the position of the stars in the sky, figured out (despite the Sun drowning out the light of the other stars) where the Sun was every day, and noticed that every 365 days or so the positions would overlap and start going over the same locations again. That line marked a loop around the celestial sphere. Astronomers marked the line using twelve zodiac constellations positioned near and through the loop.

What is the **difference** between the **ecliptic plane** and Earth's **equatorial plane**?

The equatorial plane is the plane of Earth's equator extended indefinitely out into space. It turns out that Earth's rotation around its axis is not lined up with the ecliptic plane. Instead, Earth is tilted about 23.5 degrees. This tilt is the main cause of the seasons on Earth.

How does the **motion of Earth around the Sun** cause the seasons to occur?

Some people mistakenly think that the seasons are caused by Earth being farther from the Sun in winter and closer to the Sun in summer. This is incorrect; Earth's elliptical orbit is close enough to a perfect circle that distance is not the reason. In fact, Earth is closest to the Sun in early January and farthest in early July, which is exactly the opposite of our summer and winter seasons. The reason for the seasons has to do with the angle at which sunlight strikes any particular place on Earth at any given time of year. The angle changes throughout the year because the tilt of Earth's axis differs from the ecliptic. Since the Earth is tilted 23.5 degrees, the Sun's rays hit the northern and southern hemispheres unequally. When the Sun's rays hit one hemisphere directly, the other hemisphere receives diffused rays. The hemisphere that receives the direct rays of the Sun experiences summer; the hemisphere that receives the diffused rays experiences winter. Thus, when it is summer in North America, it is winter in most of South America, and vice versa.

Is the **Earth's rotation slowing down**?

Yes. About 400 million years ago, there were 400 days a year, versus the present day 365. Eventually, if the Sun doesn't die first, the Earth will stop rotating completely.

Does the **Earth's tilt** ever **change**?

Yes. Our planet actually wobbles a bit, like a spinning top running out of steam. Currently, the "axial tilt" of our planet is about 23.5 degrees, which is somewhere in the middle of its total capable range of 22.1 to 24.5 degrees. The change in tilt occurs over a period of about 41,000 years.

What is **precession**?

Precession is a phenomenon that results from the planet's changing tilt. You can think of it as a kind of wobbling effect. About 12,900 years from now, the North Pole will be tilted toward the Sun in January and away from it in June. This means that the winter season in the North will occur during the months that are now considered summer (late June through early September) and summer will occur from January through March. This change will be gradual over time, and no one alive today or for many generations will be aware of it.

What is **orbital inclination**?

Not only does our planet tilt back and forth and wobble while it's doing it, but it also ranges up and down relative to the invariable plane (the plane, in simplified terms, passing through the solar system's center of mass). If you imagine the Earth's orbit as forming a disk like a CD, then imagine the CD wobbling back and forth instead of spinning on a level plane (formed by the invariable plane), then you might get the idea of orbital inclination. The current orbital inclination of the Earth causes it to pass through the invariable plane in early January and early July. The invariable plane carries with it more space dust and debris than is found above and below this plane; thus, as the Earth passes through the invariable plane, the atmosphere comes in contact with more space dust, which means we see more meteor showers and meteorites. The space dust also contributes to cloud formation in the upper atmosphere: noctilucent clouds.

What is the difference between **perihelion** and **aphelion**?

Perihelion is the point where the Earth is closest to the Sun (91.4 million miles, or 147 million kilometers). This occurs around January 3 every year. Aphelion is when

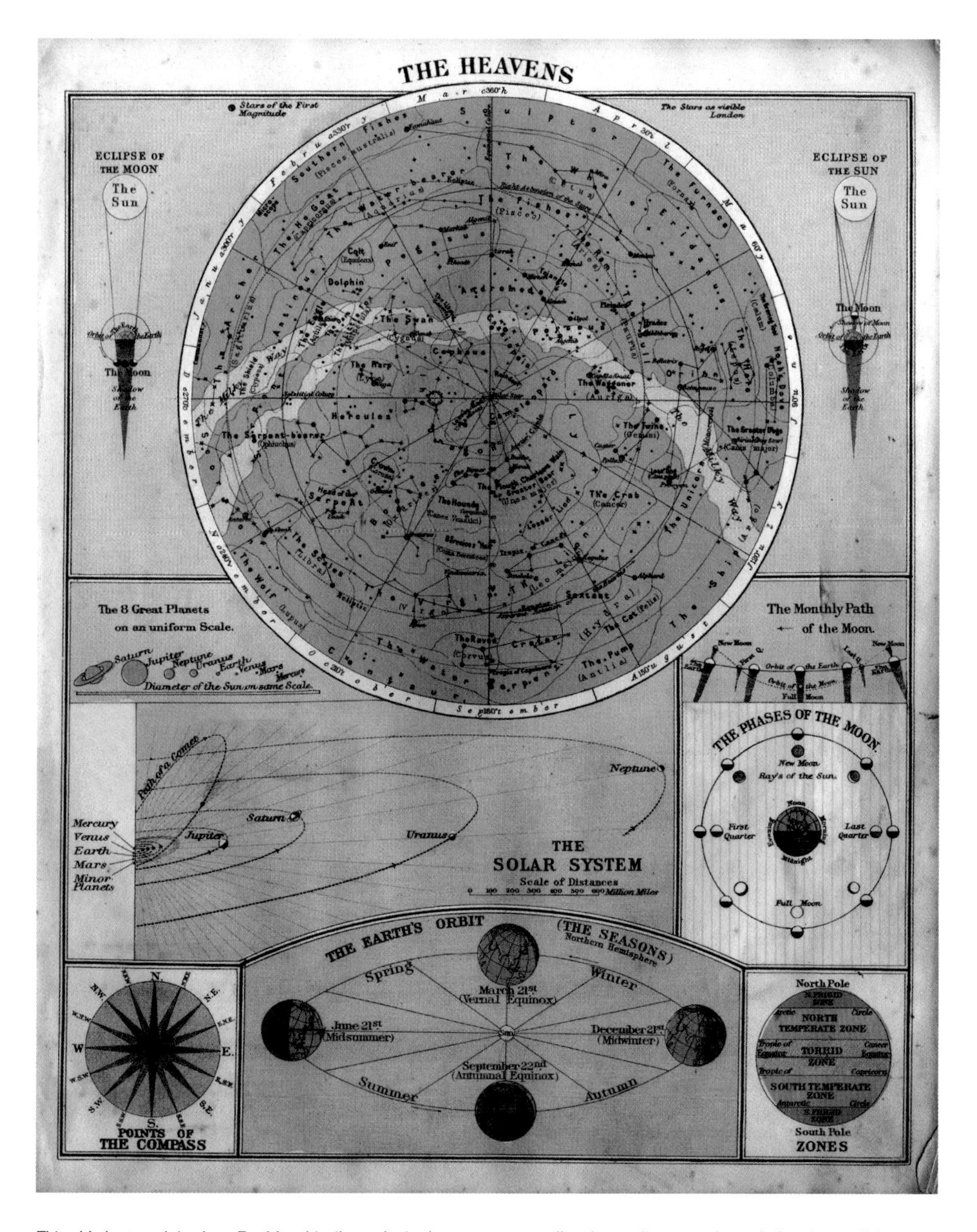

This old chart explains how Earth's orbit tilt results in the seasons, as well as how eclipses are formed, the phases of the Moon, and the latitudinal zones.

our planet reaches its farthest point from the Sun (94.5 million miles, or 152 million kilometers) around July 4. This variation does not have much effect on weather patterns or seasons.

What are **solstices** and when do they occur?

A solstice is a time of the year when Earth is pointed either the closest toward the Sun or the farthest away from it. On the summer solstice, there are more minutes

Archeologists generally agree that ancient Stonehenge near Wiltshire, England, was used long ago to mark solstices and equinoxes.

of daylight than there are on any other day of the year; on the winter solstice, there are fewer minutes of daylight than there are on any other day of the year. In the northern hemisphere, the summer solstice occurs around June 21 of each year, when the North Pole is pointed closest toward the Sun, and the winter solstice occurs around December 21 of each year, when the North Pole is pointed farthest away from the Sun.

What are **equinoxes** and when do they happen?

An equinox is a time of the year when, in the course of Earth's orbit, our planet is at a location where the equatorial plane and the ecliptic plane intersect. In other words, the tilt of Earth's axis is pointed perpendicular to the line between Earth and the Sun at an equinox—Earth's poles are tilted neither "toward" nor "away" from the Sun, but tilted off to the "side." On the day of an equinox, there are as many minutes of daylight as there are of night—hence the term "equinox," meaning "equal darkness." In the northern hemisphere, the vernal (spring) equinox occurs around March 21 of each year, and the autumnal (fall) equinox occurs around September 21.

Do **large-scale weather patterns** lead to **season trends**?

Generally, no. One might believe, for example, that a milder-than-normal winter might be followed by a warmer-than-usual spring and summer. Actually, meteorologists have found no such reliable patterns. In fact, many times a warm winter is followed by a cold spring, or vice versa. A good example of this is the winter of 1994 to

> **Can you stand an egg on end only on the spring equinox?**
>
> It is a common legend that an egg can be balanced on its end only on the spring equinox (March 21). Actually, there's nothing magical about gravity on the spring equinox that would allow an egg to stand on end—it can happen at any time of the year with patience and perseverance.

1995. In the northern United States that season, there was a lot less snow and ice, and urban areas such as Minneapolis-St. Paul, Minnesota, saved lots of money on road salt. However, the following spring was decidedly colder, and Minnesotans saw ice-covered lakes and ponds well into the month of May. Looking back farther in history, the Dust Bowl years of the 1930s saw severe extremes, with the United States experiencing many of its all-time record lows and highs in 1933, 1934, 1936, and 1937.

What are the **"dog days" of summer**?

The "dog days" of summer comprise a period of extremely hot, humid, and sultry weather that traditionally occurs in the northern hemisphere in July and August (traditionally, the days run from July 3 through August 11). The term comes from the dog star, Sirius, in the constellation Canis Major. At this time of year, Sirius, the brightest visible star in the sky, rises in the east at the same time as the Sun. Ancient Egyptians believed that the heat of this brilliant star added to the Sun's heat to create hotter weather. Sirius was blamed for everything from the withering droughts to sickness to the discomfort that occurred during this time.

What are **halcyon days**?

This term is often used to refer to a time of peace or prosperity. Among sailors, it is the two-week period of calm weather before and after the shortest day of the year, approximately December 21. The phrase is taken from halcyon, the name the ancient Greeks gave to the kingfisher. According to legend, the halcyon built its nest on the surface of the ocean and was able to quiet the winds while its eggs were hatching.

What is **Indian summer**?

The term Indian summer dates back to at least 1778 and may relate to the way Native Americans availed themselves of the nice weather to increase their winter food supplies. It refers to a period of pleasant, dry, warm days from middle to late autumn that usually occur after the first killing frost.

What is a **January thaw**?

Mostly seen in the northeastern United States and in the United Kingdom, a January thaw is a brief mid-winter period—usually late in the month—in which temperatures moderate somewhat. The Midwest can also experience such thaws, occasionally with startling changes in temperature. For example, in January 1992, northwestern Iowa had a January thaw in which temperatures rose from –60°F (–51°C) to above freezing in just two weeks. While the change was a welcome one for many people, the thaw sadly melted the giant ice palace sculpture that had been on display for the Saint Paul Winter Carnival.

How should one **prepare** when the weather forecast calls for the season's **first freeze**?

If you own a home, there are several things to do to prepare for the onset of winter. Make sure that your furnace is in good working order and that you have clean air filters. If you have a chimney and a wood-burning fireplace, have a professional chimney sweep clear it of inflammable creosote, a fire hazard responsible for many house fires annually in the United States. Also, check the chimney outside for any birds' nests, which are also a potential fire hazard. Outside, in the yard, drain garden hoses and check the sprinklers to make sure they are clear. Frozen hoses and sprinkler systems can cause pipes to burst.

ATMOSPHERE BASICS

How high up does the **atmosphere reach**?

The end of the atmosphere is not like the horizon, where you can definitely say, "This is where the Earth ends and the atmosphere begins." Rather, as one travels higher and higher, the atmosphere gets thinner and thinner. One can say, for practical purposes, that the upper atmosphere begins to be indistinguishable from outer space at about 435 miles (700 kilometers) altitude, but that is really just a random place to draw the borderline. The density of the atmosphere is getting very thin indeed at an elevation of 370 miles (about 600 kilometers). At this height, there are about six miles (10 kilometers) between each molecule (this gap is known as the "mean free path." The air pressure here is, effectively, zero.

How did **Earth's atmosphere form**?

Some of Earth's atmosphere was probably gas captured from the solar nebula four and a half billion years ago, when our planet was forming. It is thought that most of Earth's atmosphere was trapped beneath Earth's surface, escaping through volcanic eruptions and other crustal cracks and fissures. Water vapor was the most plentiful gas to spew out, and it condensed to form the oceans, lakes, and other surface water. Carbon dioxide was probably the next most plentiful gas, and much of it dissolved in the water or combined chemically with rocks on the surface. Nitrogen came out in smaller amounts, but did not undergo significant condensation or chemical reactions. This is why scientists think it is the most abundant gas in our atmosphere.

The high concentration of oxygen in our atmosphere is very unusual for planets, because oxygen is highly reactive and combines easily with other elements. In order to maintain oxygen in gaseous form, it must constantly be replenished. On

Where does the word "gas" come from?

The person who is credited with coming up with the word "gas" is Flemish physician Jan Baptista van Helmont (1577–1644). His experiments with gas and volume taught him that gases always take up all the space within a container (they do not leave a vacuum). Thus, he surmised that substances in a gaseous state exist in a chaotic form. The Flemish pronunciation for "chaos" sounds like "gas," and thus the word was born.

Earth, this is accomplished by plants and algae that conduct photosynthesis, removing carbon dioxide from the atmosphere and adding oxygen into it. It was during the Carboniferous Period about 300 million years ago that plant growth dramatically changed the atmosphere, increasing the amount of oxygen to 35 percent. Today, the oxygen content is not quite as rich (21 percent).

How is **Earth's atmosphere** important to **life** on Earth?

Very few life forms on Earth can survive for any length of time at all without Earth's atmosphere. We breathe the atmosphere; and it blocks harmful radiation from space. The pressure it provides keeps surface water liquid, and the greenhouse effect it produces keeps us warm.

How **thick** is **Earth's atmosphere**?

Earth's atmosphere extends hundreds of miles beyond its surface, but it is much denser at the surface than at high altitudes. About half of the gas in Earth's atmosphere is within a few kilometers of the surface, and 95 percent of the gas is found within 12 miles (19 kilometers) of the surface.

Are we **losing** our **atmosphere**?

Yes, but don't worry; the number of molecules and ions escaping our atmosphere is very tiny and will not deplete our atmosphere significantly for billions of years. Scientists monitoring the magnetosphere learned that periodic changes in the magnetosphere help to accelerate particles, especially ions, with enough speed to escape Earth's gravity. If the Earth's gravity were weaker, however, this interaction might have caused our planet to lose its atmosphere at a significant rate. Indeed, some astronomers speculate that this may be what happened to Mars's atmosphere.

Why is the **sky blue**?

While this might seem like a simple question to answer, it has puzzled parents of curious children for ages. The answer can be a bit long-winded. The Earth's atmosphere is made up of gases and a scattering of water and solid particles. As light from the Sun

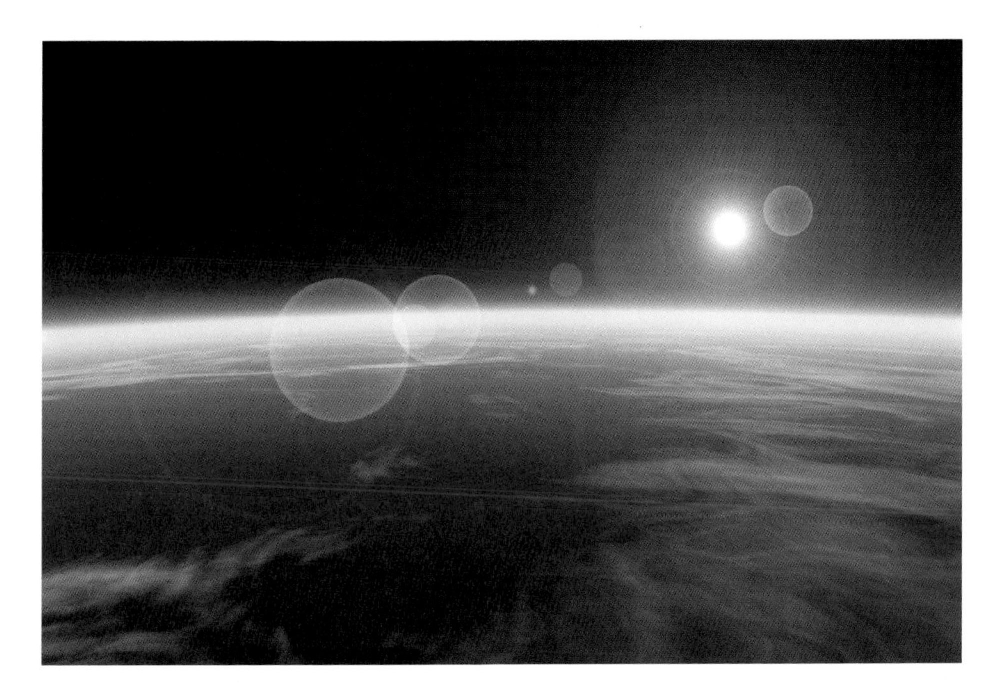

Earth's precious atmosphere formed over billions of years.

enters the atmosphere, most of it passes straight through the air, but some of it disperses because of something called Rayleigh scattering. Shorter wavelengths of light waves in sunlight (those on the blue end of the scale) are absorbed by gas molecules and then released at different angles. Because the bluer wavelengths are scattered in this manner, this is the end of the spectrum of light that reaches people's eyes. However, as your eye looks more and more toward the horizon, you are looking through a much thicker layer of air and less blue light reaches your eyes, which is why the sky appears bluer as you look up, but paler as you look toward the horizon.

At what point does the **sky turn from blue to black**?

The blueness of the sky fades away as you get higher in the atmosphere. By the time you are in the lower troposphere, such as when flying in the upper elevations in a commercial jet plane—around 35,000 feet (about 10,600 meters)—the air is quite thin and begins to look dark. Above 150,000 feet (about 45,750 meters) the sky turns increasingly black as you enter the stratosphere.

Where is the **horizon**?

Depending on your elevation above sea level—and assuming no objects such as mountains are obscuring your view and the sky is perfectly clear—the horizon appears at different distances. To calculate how far the visible horizon is, first measure the distance between the ground and your eye level. Add to this the measurement of how high your elevation is. If your total is in feet, multiply by 1.5 and then take the square root of the result (which will be in miles); if you used meters as your measuring unit, multiply by 13 and then take the square root (the result will be in kilometers).

How **far** is it possible to **see through the atmosphere**?

On a clear day that is free of haze and pollution, it is possible for people to see objects as far as 200 miles (322 kilometers) away. At night, bright lights can be visible from as far away as 500 miles (about 800 kilometers).

What are the different **scales of atmospheric motion**?

Meteorologists divide weather patterns and motions based on actions that are occurring at various scales of size. Just as economists have macroeconomics (how the economy functions on a regional, national, or international level) and microeconomics (financial concepts applied to a single household or business), meteorologists find it convenient to divide weather phenomena this way, too. Below is an explanation of the divisions.

- The macroscale (or synoptic scale) refers to weather processes affecting large areas, such as pressure systems, fronts, and the jet stream.
- The microscale concerns highly localized events, such as a tornado, a fog bank, or a small rain shower that affects an area of only a few hundred square feet or meters.
- The mesoscale deals with weather events somewhere between the micro and macro levels, usually ranging over several miles to a hundred miles or so. Thunderstorms, cloud systems, and breeze fronts are some examples of weather falling into this category. The mesoscale is sometimes divided even further into 1) meso-gamma (covering about 1 to 12 miles [2 to 20 kilometers]), 2) meso-beta (25 to 125 miles [20 to 200 kilometers]), and 3) meso-alpha (125 to 1,250 miles [200 to 2,000 kilometers]).

AIR AND AIR PRESSURE

What is **air**?

Sometimes when people think of "air" the gas that comes to mind is oxygen. Actually, oxygen molecules (O_2) are in the minority in our atmosphere. Only 21 percent of Earth's atmosphere is oxygen, with the majority of it being nitrogen gas (78 percent). The rest is a mixture of argon gas (0.9 percent), 0.035 percent carbon dioxide (CO_2), water vapor, and traces of helium, xenon, methane (CH_4), nitrous oxide (N_2O), neon, and krypton (not to be confused with kryptonite, the stuff that hurts Superman), and a scattering of dust, pollen, and other particulates.

Is there actual **salt** in **sea air**?

Yes. The air above the sea and along the coast does contain salt. On average, the concentration of salt in sea air is about 3.5 million parts per cubic foot (100,000

Pollen for a wide variety of plants fills our atmosphere, sometimes causing allergic reactions in people.

parts per cubic meter). Salt air can penetrate inland up to thousands of miles, depending on wind and air pressure conditions, and can aid in the formation of clouds because the salt particles serve as nuclei around which drops of precipitation can form. Haze along seashores is also the result of salt particles, with haze droplets typically forming when the humidity level reaches 75 percent or more.

How much **pollen** is in our air?

Pollen from plants just within American borders produce about two billion pounds of particulates every year, or about seven pounds per person.

Are the **gases** in the atmosphere **evenly distributed**?

You are not likely to stroll down a street and encounter a suddenly high concentration of oxygen or a pocket of unmixed argon gas. The constant motion of the weather due to fronts, pressure changes, varying temperatures, storms, and so on, is like putting the atmosphere in a food processor and hitting the "blend" button, and never turning it off. The percentage of each gas, therefore, will be constant anywhere below an elevation of 50 or 60 miles (80 to 95 kilometers).

How much **pressure** does the atmosphere exert upon us?

Average air pressure is 14.7 pounds per square inch (1.03 kilograms per square centimeter) at sea level. Measured in inches of mercury, this comes to 29.92 inches, or

1,013 millibars. To put it another way, a cubic yard of sea-level air weighs a bit over two pounds (about 0.7 kilograms per cubic meter).

Who introduced use of the **millibar** to measure air pressure?

English meteorologist William Napier Shaw (1854–1945) was one of the leading scientists in his field and former director of Britain's Meteorological Office from 1905 to 1920. He suggested that air pressure be measured in millibars in 1909, but it was not adopted as an international standard until 1929.

Does **air pressure change** as elevation rises above sea level?

Yes, it does. The higher you go, the less air (or atmospheric) pressure there is. Air pressure is also involved in weather systems. Closer to the ground, air pressure decreases at a rate of about 0.01 inches of mercury for every 10 feet (3 meters). By the time you are at an elevation of 18,000 feet (5,500 meters), air pressure is about half what it is at sea level. A low-pressure system is more likely to bring rain and bad weather versus a high-pressure system, which is usually drier.

What is **Gay-Lussac's Law** and why is it important in meteorology?

Joseph Louis Gay-Lussac (1778–1850) was a French physicist and chemist best known for two laws of physics about gases. One of these laws states that, in chemical reactions, gases combine in simple ratios according to volume. For example, it takes two carbon monoxide molecules (CO) to combine with one oxygen molecule (O_2) to create carbon dioxide (CO_2). This became known as Gay-Lussac's Law and is important for understanding chemical reactions of gases within our atmosphere. Gay-Lussac also published a law about gas expansion by volume occurring linearly as a function of temperature. Sometimes credited as another of Gay-Lussac's laws, it is more correctly called Charles' Law (and Gay-Lussac was one of those who gave credit where it was due). It was discovered by another French physicist, Jacques Alexandre César Charles (1746–1823), who was also a mathematician. Charles found that gases such as oxygen and nitrogen increased their volume by 1/273 for every 1.8°F (1°C) increase in temperature. From this, he extrapolated the possibility that, at absolute zero (–273°C, or –459.4°F) the volume of a gas would also be zero. Both scientists were also balloonists, which was handy, given their interest in the atmosphere.

What is **Dalton's Law**?

English meteorologist John Dalton (1766–1844) discovered that adding up the partial pressures of each gas in a gaseous mixture will equal the total pressure exerted by the mix of gases. Dalton's Law was formulated in 1801.

How much does the **entire atmosphere weigh**?

If you were able to take all the gases in the air and place them on a scale, the total weight would be 5.1 quadrillion tons (5.1×10^{15} tons), or 4.63 quadrillion metric tons.

What is the air pressure on top of Mt. Everest?

Mountain climbers ascending the peak of Mt. Everest (29,029 feet; 8,848 meters) will find the air pressure to be about a third of what they are used to. That's important not so much because of the pressure itself, but because it means there is also two-thirds less oxygen to breathe. When climbers reach 26,246 feet (8,000 meters) they are at the point above which is the "Death Zone." Many people carry oxygen tanks with them when they reach this height, although some consider it a particular test of their mettle to go without. "High altitude sickness"—hypoxia—causes fatigue, distorted vision, confusion, and loss of memory and appetite. Cerebral and pulmonary edema will set in if a person goes too long before returning to normal air pressure, and this can prove fatal within a couple of days. Frigid temperatures and lack of oxygen have killed dozens of people who have attempted to reach the top of the world. Overall, more than 150 climbers have died on the world's tallest mountain.

Who showed that **air pressure** decreases as **altitude** increases?

French physicist and mathematician Blaise Pascal (1623–1662) was inspired by fellow physicist Evangelista Torricelli (1608–1647) to test the idea that the air in the atmosphere is much like seawater in an ocean. Since pressure in an ocean or lake increases as one descends into the depths, Pascal hypothesized that air pressure in a valley would be higher than on top of a mountain. To test the idea, in 1646 he asked his brother-in-law, Florin Perier (1605–1672), to use a barometer (a new invention at the time) and measure the pressure both at the top of the French volcanic peak Puy de Dôme and in the village of Clermont-Ferrand in Auvergne. The difference in elevation between the two is about 3,900 feet (1,200 meters). Taking along witnesses to verify his readings, Perier found that the air pressure at Clermont-Ferrand was 28 inches of mercury and at the top of Puy de Dôme it was only 24.6 inches. A monument to Pascal was later erected in Clermont-Ferrand to honor this achievement, and there is also a meteorological observatory atop Puy de Dôme.

Is air really **"thin"**?

Like anything else, "thin" is a relative term. Compared to the vacuum of space, air is very dense, but compared to a chunk of marble, or even a bottle of water, it is very thin indeed. For something in a gaseous state, air is actually quite thick. At sea level, molecules in the air are spaced only about one millionth of an inch apart.

What is a **front**?

A front is any boundary existing between differing masses of air; that is, masses that have different overall temperatures and humidity. Fronts come in four types: warm, cold, stationary, and occluded. A warm front is a front that is advancing over a cool-

Why do the Denver Broncos have a particular advantage at Invesco Field at Mile High?

After the original Mile High Stadium was demolished in 2001, the Denver Broncos moved to Invesco Field to play football. Here, as with other teams in Denver, such as the Rockies, players have a distinct advantage over visiting teams because their lungs are already accustomed to the higher altitude. Fortunately, the human body can quickly adapt, and if a visiting athlete is able to arrive in the Denver area a couple days before a game, he or she should be able to play on a par with the locals.

er mass of air, while a cold front is just the opposite. Stationary fronts, as one might guess, represent boundaries between warm and cold air that are in relative equilibrium, although they can still move back and forth by as much as several hundred miles and yet be considered "stationary."

An occluded front occurs whenever a cold front, instead of merely overtaking warm air, actually separates and breaks apart warm mass of air. Occluded fronts come in both warm and cold varieties. In a warm occluded front, the cold air at the advancing side of the warm front is *cooler* than the air in the cold front that is overtaking it. In a cold occluded front, the cold air at the advancing side of the warm front is *warmer* than the air in the advancing cold front.

What is a **dry front**?

A dry front—also called a dry line or dew point front—is a borderline separating a mass of dry air from one of much more humid air. Often found east of the Rocky Mountains, these fronts will find the warmer, drier air lifting the cooler, more humid air ahead of it in the higher altitudes, while humid air near the ground is denser than the dry air and the drier air will flow over it. The result is an air mass reversal that can precipitate the formation of cumulonimbus clouds, thunderstorms, and, quite often, tornadoes.

What is a **barometer**?

A barometer is a device that measures air pressure. A standard barometer consists of a glass tube filled with mercury (a liquid metal) that is inserted into a reservoir, which also contains mercury. When the surrounding air pressure exerts more weight on the reservoir than the mercury in the tube does, the mercury level rises, and vice versa.

What is **barometric pressure** and what does it mean?

Barometric, or atmospheric, pressure is the force exerted on a surface by the weight of the air above that surface, as measured by an instrument called a barometer.

Some weather fronts are so well defined that anyone can see them coming.

Pressure is greater at lower levels because the air's molecules are squeezed under the weight of the air above. So while the average air pressure at sea level is 14.7 pounds per square inch, at 1,000 feet (304 meters) above sea level, the pressure drops to 14.1 pounds per square inch, and at 18,000 feet (5,486 meters) the pressure is 7.3 pounds, about half of the figure at sea level. Changes in air pressure bring weather changes. High pressure areas bring clear skies and fair weather; low pressure areas bring wet or stormy weather. Areas of very low pressure have serious storms, such as hurricanes.

Who **invented** the **barometer**?

Invented in 1644 by Evangelista Torricelli (1608–1647), a barometer is a device for measuring air pressure. Torricelli was a student (for a brief three months) of Galileo Galilei (1564–1642), and he was inspired by his mentor's observation that piston pumps can only lift water up 33 feet (about 10 meters), after which point it is impossible to pump the water any higher. After Galileo died, Torricelli continued to build on this observation. He theorized, correctly, that air had weight and, therefore, exerted pressure. He tested his theory by filling a dish with mercury (he used mercury because it was denser than water and therefore would require a much smaller amount to indicate pressure changes). He then took a four-foot-long glass tube that was open on one end, filled it with mercury, and turned it upside down with the open end beneath the surface of the mercury. Some, but not all, of the mercury exited the tube; 30 inches (760 millimeters) remained. This meant that the remaining mercury in the tube stayed in the tube because air in the atmosphere was

What was the "Tempest Prognosticator"?

In 1851 Dr. George Merryweather displayed his "Tempest Prognosticator" at London's Great Exhibition. Remarkably, this "leech barometer" used the infamous blood-sucking worms to predict changes in air pressure and, thus, forecast inclement weather. Leeches require moisture to stay alive, and so they usually remain underwater unless it is rainy and wet outside their river and pond homes. It is believed that leeches possess a kind of natural barometer that tells them when a low pressure system is bringing wet weather and they can venture outside to seek victims to feed upon. Merryweather's contraption worked by placing one leech in each of a series of twelve bottles filled with about an inch and a half of water. The bottles were arranged in a circle, and at the top of each bottle was a wire connected to a bell. When the air pressure was high, the leeches would remain in the water. When the air pressure began to fall, however, the leeches would become more active and climb up the bottles, eventually ringing the bells.

exerting pressure on the surface of the mercury in the dish. Not only did this experiment prove Torricelli's theory that air had pressure, but he was also the first to create a vacuum (now called a Torricellian vacuum).

The word "barometer," which means "weight measure," was not coined until 1665 by Irish scientist and theologian Robert Boyle (1627–1691). Boyle came up with a new design for the barometer in which a U-shaped tube was used, eliminating the need for a mercury reservoir. English physicist Robert Hooke (1635–1703) made another improvement on the barometer by creating an easy-to-read dial display.

What is an **aneroid barometer**?

The word "aneroid" means "without fluid," and so aneroid barometers do not need mercury in order to work. French inventor Lucien Vidie (1805–1866) built upon a concept first proposed by German mathematician Gottfried Leibniz (1646–1716) in which a metallic capsule surrounded by a vacuum could be used to measure air pressure. Using very thin pieces of metal, Vidie managed to connect such a capsule to highly sensitive dials displayed behind glass within an encasement. This was highly detailed work on the level of the finest clock craftsmanship. Aneroid barometers were very difficult to make in Vidie's time, but high-tech instruments are produced today using such devices as electron beams welding copper beryllium alloys. Because aneroid barometers are made of metal, they are also sensitive to changes in temperature and altitude. Bimetallic strips can be used to compensate for temperature, but altitude poses more of a problem. For this reason, aneroid barometers work best at elevations below 3,000 feet (about 915 meters), but they can be calibrated for higher altitudes if needed.

Barometers come in a variety of types, but they are all used for the same purpose: to measure air pressure.

What is a **digital barometer**?

A digital barometer is an aneroid barometer that works by running an electrical current between two strips of metal. The current measures the distance between these strips (which are affected by air pressure) and translates these into an electronic display.

What is a **banjo barometer**?

The banjo barometer is a barometer set into a banjo-shaped case. It was developed by Robert Hooke (1635–1703) and was a very popular design because the large dial was easy to read and was large enough so that you could get very detailed readings.

What is **sea-level air pressure**?

The air pressure at sea level averages 29.92 inches of mercury, or 1,013.25 millibars (mb). No matter what their elevations, weather stations calculate barometric pressures in their area according to what they would be at sea level. In this way, reports remain consistent everywhere.

How do you **convert** inches of **mercury to millibars**?

When measuring air pressure, sometimes inches of mercury are used, and sometimes millibars. To convert inches of mercury to millibars, multiply the number of inches by 33.8637526 (or, 33.86 will give you an accurate enough measurement).

A barograph resembles the more familiar seismograph, but instead of tracking earthquakes it draws lines on paper indicating changes in air pressure.

To do the calculation in reverse, multiply millibars by 0.0295301 (or 0.03 for an estimate) to give you inches of mercury.

What is a **hectopascal**?

A hectopascal (hPa) is the same thing as a millibar. Some meteorologists, especially outside the United States, use hectopascals instead of millibars or inches of mercury because the hPa is the International System of Units (SI) standard.

What are the **most extreme barometer readings** ever measured?

On December 29, 2004, a barometer reading of 32.25 inches of mercury was taken at Tonsontsengel, Mongolia—a world record. At the other extreme, a reading of 25.63 inches was recorded on October 12, 1979, in the Philippines within the eye of Typhoon Tip.

What is a **barograph**?

Resembling a seismograph, a barograph records changes in air pressure over time. A recording arm with a pen on one end moves side to side, drawing a line over a paper or foil chart that moves over a rotating barrel.

LAYERS OF THE ATMOSPHERE

How many **layers** does the **Earth's atmosphere** contain?

The atmosphere, or "skin" of gas that surrounds the Earth, consists of six layers that are differentiated by temperature:

1. *The troposphere* is the lowest level. It averages about 7 miles (11 kilometers) in thickness, varying from 5 miles (8 kilometers) at the poles to 10 miles (16 kilometers) at the equator. Most clouds and weather form in this layer. Temperature decreases with altitude in the troposphere.
2. *The stratosphere* ranges between 7 and 30 miles (11 and 48 kilometers) above Earth's surface. The ozone layer, which is important because it absorbs most of the Sun's harmful ultraviolet radiation, is located in this band. Temperatures rise slightly with altitude to a maximum of about 32°F (0°C).
3. *The mesosphere* (above the stratosphere) extends from 30 to 55 miles (48 to 85 kilometers) above the Earth. Temperatures here decrease with altitude to –130°F (–90°C).
4. *The thermosphere* (also known as the hetereosphere) is between 55 to 435 miles (85 to 700 kilometers) above Earth's surface. Temperatures in this layer range to 2,696°F (1,475°C).
5. *The ionosphere* is a region of the atmosphere that overlaps the others, reaching from 40 to 250 miles (65 to 400 kilometers). In this region, the air becomes ionized (electrified) from the Sun's ultraviolet rays. It is divided into three subregions: 1) the D Region (40 to 55 miles [65 to 90 kilometers]); 2) the E Region (also called the Kennelly-Heaviside layer) at 56 to 93 miles (90 to 150 kilometers); and 3) the F Region (93 to 248 miles [150 to 400 kilometers]), which is further separated into the F_1 layer and the F_2 layer (also called the Appleton layer), with the dividing line being at about 150 miles (240 kilometers) above sea level.
6. *The exosphere* lies above the thermosphere and includes everything above 435 miles (700 kilometers) high. In this layer, temperature no longer has any meaning.

How is the **troposphere defined**?

The troposphere is considered as the layer closest to Earth and is also the region where temperatures reliably decrease with altitude. The troposphere is thickest at the Earth's equator, reaching heights of about 11 miles (17 to 18 kilometers), and this is, therefore, also where you will find the coldest tropospheric temperatures. It might seem rather ironic that, right above the world's steamiest tropical forests, temperatures can be as low as –110°F (–79°C).

Who figured out that **temperatures rise in the stratosphere**?

Pioneering French meteorologist Léon Philippe Teisserenc de Bort (1855–1913) conducted an experiment using helium balloons and temperature sensors. He

A circa 1900 photograph of Teisserenc de Bort (left), standing with Blue Hill Meteorological Observatory founder and fellow atmosphere researcher Abbott Lawrence Rotch (1861–1912). *(NOAA)*

learned that, after about seven miles (11 kilometers), the air stopped becoming cooler and leveled off for as high as the balloons could go. He concluded that the atmosphere was divided into two layers, which he named the troposphere and the stratosphere. Later, in the 1920s, meteorologists Gordon Miller Bourne Dobson (1889–1976) and F.A. Lindemann, First Viscount Cherwell (1886–1957), used studies of meteor trails to learn that temperatures warmed in the atmosphere as high as 30 miles (48 kilometers) up. Dobson concluded that ultraviolet radiation absorbed by the ozone in the stratosphere was the reason for the warmer air.

What **mineral** were scientists surprised to see in **high quantities** in the **stratosphere**?

Scientists discovered a higher-than-expected amount of salt in the mesosphere within the stratosphere. The current theory is that the salt has been left behind by meteor activity.

What is the **tropopause**?

The tropopause is the layer between the troposphere and the stratosphere that hovers around 10 miles (16 kilometers) above the ground. Within the tropopause is the tropopause break, a region through which water vapor and air can easily pass from the troposphere to the stratosphere.

How is the **ionosphere** important in the transmission of **radio waves**?

When ultraviolet light enters the atmosphere, it ionizes atoms in the ionosphere through a process called photoionization, which releases free electrons into this region of the atmosphere. It is these free electrons that make radio wave transmissions possible. Depending on the frequency of the radio waves, transmissions travel for shorter or longer distances. Lower-frequency waves bounce off the ionosphere at a lower elevation and thus travel a shorter distance than higher frequency waves. Very high frequency waves are used when communicating with satellites or anything out in space because they can completely escape the atmosphere.

British physicist Oliver Heaviside (1850–1925) and American electrical engineer Arthur Edwin Kennelly (1861–1939) independently theorized the existence of an

Why are radio transmissions weaker at night than during the day?

At night, of course, much less light is entering the atmosphere, which means that less ionization is occurring and fewer electrons are available for radio waves to bounce off of. The result is that radio transmissions are weaker.

ionosphere and that certain wave frequencies would bounce off it and be reflected back to Earth. It was radio pioneer Guglielmo Marconi (1874–1937) who first took advantage of this theory to conduct the first transmission from Cornwall, England, to Newfoundland in 1901. The waves became known as radio waves, and Marconi is credited as the inventor of the radio. The E region of the ionosphere was named after Kennelly and Heviside to honor their work.

What is an **ionospheric storm**?

When a coronal mass ejection (solar flare) occurs on the Sun's surface, it can dramatically increase the amount of photoionization in the ionosphere. An overwhelming amount of free electrons in the upper atmosphere results, and this can cause problems with radio communications.

What is the **mesosphere**?

The mesosphere is the uppermost layer of the stratosphere. Below the mesosphere, at altitudes of 25 to 40 miles (40 to 65 kilometers), is a warm layer of the stratosphere that contains a high concentration of ozone molecules that block ultraviolet light.

Why is the **thermosphere so hot**?

While the temperature of the thermosphere can exceed 3,600°F (1,982°C), the atmosphere here is so thin that an ordinary thermometer would not register this temperature (a regular thermometer would indicate temperatures below freezing). Instead, special instruments are used to measure the speed of the few particles that *are* in the thermosphere, and these indicate the extraordinary highs. Atoms and molecules in the thermosphere are excited by radiation from outer space.

THE OZONE LAYER

What is the **ozone layer**?

The ozone layer is part of the stratosphere, a layer of the Earth's atmosphere that lies about 10 to 30 miles (16 to 48 kilometers) above the surface of the Earth. Ozone (O_3) is like regular gaseous oxygen (O_2) with an extra oxygen atom attached to it. It

Before the ozone layer was discovered, someone must have discovered ozone. Who was that?

Dutch chemist Martinus van Marum (1750–1837) discovered ozone around the year 1785, while conducting experiments with electricity. As many high school science students now do, van Marum smelled a distinctive odor while working with oxygen and electricity. Van Marum, who is also credited with discovering carbon monoxide, did not, however, identify the source of the odor as a unique gas molecule. It was not until 1840 that German-Swiss chemist Christian Friedrich Schönbein (1799–1868) correctly identified ozone as a gas, which he named after the Greek *ozein,* meaning "smell." Finally, Swiss chemist J.L. Soret worked out its chemical structure as being a molecule consisting of three bonded oxygen atoms (O_3).

is created when short-wavelength ultraviolet radiation interacts with O_2 molecules. The energy from the radiation breaks the molecules apart, which then recombine into ozone.

The ozone layer is important because it protects life on Earth from harmful ultraviolet radiation. While it does not absorb *all* of this radiation (otherwise, it would be impossible for you to get a tan!), it prevents about 80 percent of it from reaching life on Earth. As anyone who knows about melanoma can tell you, too much ultraviolet radiation can lead to cancer.

Who discovered the **ozone layer**?

In 1913, French physicists Henri Buisson (1873–1944) and Charles Fabry (1867–1945; full name, Marie Paul Auguste Charles Fabry) theorized that an ozone layer existed in the upper atmosphere. It was confirmed in a series of measurements of ultraviolet radiation levels that were recorded by W.N. Hartley and A. Cornu from 1879 to 1881.

How did **Charles Fabry** discover that the **ozone absorbs ultraviolet light**?

Fabry (1867–1945) invented the interferometer (a device measuring how light waves interfere with each other) with fellow French physicist Albert Pérot (1863–1925). He used this Fabry-Pérot interferometer to measure the Doppler effect on light in the laboratory. Then, in a 1913 experiment, he used it to learn that UV radiation is absorbed by the ozone layer.

Has there **always been an ozone layer**?

No. Before plants evolved on the planet, there was no ozone layer, because plants are responsible for converting carbon dioxide in the atmosphere into oxygen molecules. So, life on Earth actually began to evolve before there was an ozone layer.

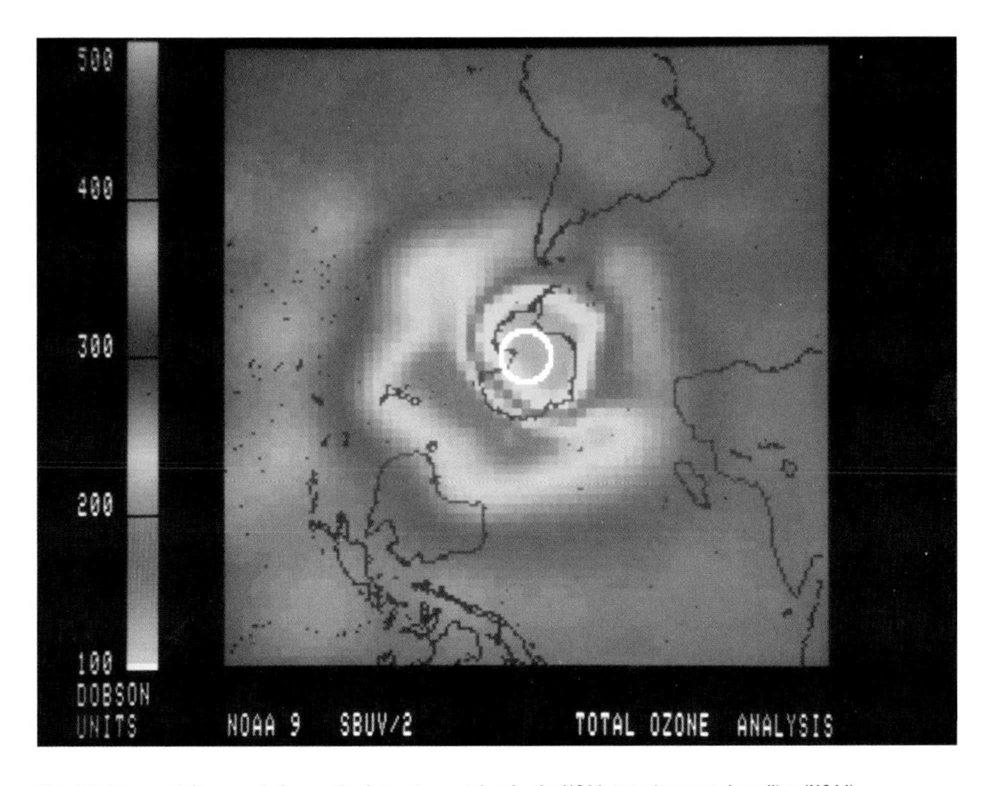

This 1987 image of the ozone hole over the Antarctic was taken by the NOAA-9 environmental satellite. (*NOAA*)

If **ozone is good** for us, why do we hear about **ozone alerts and "bad ozone"**?

Ozone is great when it is high up in the atmosphere and doing its job of protecting us from radiation, but when it is down near ground level it is toxic to those who breathe it in. Car exhaust fumes and other sources of pollution include ozone, which is seen in smog. Ozone pollution can make people sick and damage crops. Small amounts of ozone are also produced by electrical storms (if you have ever performed electricity experiments in a high school or college laboratory, you might be familiar with the ozone smell produced by even small bolts of electricity).

What are some **other effects** of **ultraviolet radiation**?

Small amounts of ultraviolet (UV) radiation can actually be good for you because it aids in the production of Vitamin D in the body. However, you only need about 10 to 15 minutes of sunlight exposure a day to get this benefit (not getting enough sunlight leads to Vitamin D depletion, which can lead to depression and other symptoms in humans and is, indeed, a chronic problem in populations located in northern and extreme-southern climates).

Besides the risks of cancer—especially melanoma—overexposure to UV radiation can cause cataracts or inflammation of the cornea ("snow blindness"). If exposure is not too long, the eye can heal itself from this inflammation, but prolonged

> ### What is the Umkehr effect?
>
> Swiss astronomer Paul Götz published a paper in 1931 in which he described how ultraviolet light is affected by the ozone layer. Ozone absorbs different wavelengths of light in the ultraviolet range at varying amounts, and these amounts change depending on the angle of the Sun in the sky ("umkehr" is German for "change" or "convert"). Measuring the difference between the amount of light being received at the two wavelengths can tell scientists how much ozone is in the atmosphere.

exposure could lead to permanent blindness; cancer of the eye is also a possibility. It is also believed that too much UV radiation weakens immune systems, though studies are still being conducted to more fully understand this health risk. Interestingly, it has also been found that, in areas where ozone levels are lower and more UV radiation penetrates to ground level, certain construction materials such as wood and some plastics degrade at a higher rate than normal.

High UV levels also, of course, affect plants and animals, though some are at a higher risk than others. Scientists have learned, for example, that soy bean crops and some types of rice could die if the ozone were too severely depleted. Also, young pine tree needles are damaged by UV light, but mature needles, which have a waxy coating, are protected. In the oceans, some forms of plankton could die or be severely depleted if the ozone was not doing its job. The result would be a breakdown in ocean food chains that could be devastating. The effects on wild animals are not well known, though nocturnal animals would likely be unaffected, and many diurnal animals have fur or feathers that protect them from radiation. However, skin around the eyes and ears are often more exposed to the sunlight, and animals would be as susceptible to eye problems as much as humans.

What is the mission of **NASA's *Aura* satellite**?

Launched on July 15, 2004, the *Aura* satellite is on a mission to monitor changes in the Earth's atmosphere, particularly the ozone layer. Instruments on the satellite measure the chemistry and dynamics of the upper atmosphere. The data it will gather will be used to predict alterations in air quality and climate change.

Why is there a **hole in the ozone layer**?

The ozone layer is not evenly distributed around the planet. It is thicker around the equator and nearby latitudes and thinner as one progresses north and south. This is true of the Earth's atmosphere in general because the planet's spin causes the planet to bulge slightly around the middle; the gravitational pull is consequently a little weaker and the atmosphere thickens. At the poles, the atmosphere is thinner, including the ozone layer. In addition, because ozone is dependent upon the interaction of sunlight and oxygen, there is naturally less ozone at the poles; furthermore, ozone layers fluctuate naturally over time due to numerous factors affecting climate and sunlight levels.

Since 1975, scientists believe that more than 33 percent of the ozone layer has disappeared. There is a seasonal factor to the reduction in ozone at any given time during the year, too. At different times, the ozone layer naturally declines or rises. But scientists also know that chlorofluorocarbons (CFCs), which are used for air conditioning, aerosol sprays, and halon in fire extinguishers, along with methane (CH_4) and nitrous oxide (NO_2), are broken down by UV radiation, freeing carbon, chlorine, and nitrogen atoms that then react with ozone molecules and destroy them. CFCs are particularly bad because they last so long in the atmosphere. One CFC molecule can destroy 100,000 molecules of ozone!

Who helped **link CFCs** to the **destruction of the ozone layer**?

Mexican atmospheric chemist Mario J. Molina (full name, José Mario Molina-Pasquel Henríquez [1943–]) and American atmospheric chemist Frank Sherwood Rowland (1927–) are generally acknowledged as the scientists who first explained how chlorofluorocarbons were destroying the ozone layer. A paper they published together in 1974 first explained how the process works about four years after scientists began to understand that ozone levels were declining in the upper atmosphere. The result of their work led the United States government to ban CFCs in aerosol cans in 1978.

How was the **ozone hole discovered**?

The famous meteorologist Gordon Miller Bourne Dobson (1889–1976) was the first to make accurate measurements of the ozone in the 1920s, but it wasn't until 1979 that the depletion of the ozone was observed at the South Pole by the *Nimbus 7* satellite. Today, a network of "Dobson spectrophotometers" have been set up around the world to monitor changes in the ozone.

Is the **hole in the ozone layer** causing frog **species extinctions**?

Biologists have known for a long time that frogs are very vulnerable to changes in their environment. Frogs across the globe were being found with deformities, such as extra legs, and species were going extinct. By the mid 1990s, it was still being speculated that the cause of the mutations was the weakened ozone layer, which was allowing too much ultraviolet radiation to filter onto the planet. Today, however, most scientists believe that the culprit is fertilizers leaking into the lakes and rivers where frogs live. The fertilizers

A *Dendrobates tinctorius,* or species of dart poison frog, is one of many frog species in danger of extinction for many reasons, some of which are related to environmental changes.

cause certain types of snail species to thrive, and these snails often host parasites. The parasites, in turn, infect frogs when they are still in their tadpole stage. Cysts form on the tadpoles, which creates the mutations that are being observed.

Besides the malformations seen in frogs, there is another, even more troubling concern: many species of frogs—some estimate about 100 species are vulnerable—are threatened with extinction, and many others have already disappeared. In this case, the culprit *is* global warming. Because frogs have thin skin, they are vulnerable to environmental changes. Increased temperatures have caused fungi—some scientists specifically blame the chytrid fungus—to infect frog skin, and this leads to the lethal disease *Batrachochythrium dendrobatidis* (BD). The good news is that it is easy to treat and cure frogs; the bad news is that, even if they are treated, once released back into the wild they are likely to be reinfected. To help arrest the extinction rate, zoos around the world have been rescuing sample populations and breeding them in captivity.

Why is there an **ozone hole** at the **South Pole** but there isn't one at the **North Pole**?

The harmful chemicals that destroy the ozone layer have to be carried up into the stratosphere by clouds in order to react with ozone. The land mass at the South Pole (Antarctica) creates the necessary weather conditions for this to happen, while the North Pole is covered in water, causing winds in the upper atmosphere to blow away pollutants. The bad news is that some scientists fear that increasing levels of pollution may result in a hole in the ozone at the North Pole in about 20 years.

How **big** is the **hole in the ozone**?

In 2007 the ozone hole was measured to be 9.3 million square miles (24 million square kilometers) in size. But this was a smaller hole than the record, set in September 2006, when the ozone hole was a gaping 10.6 million square miles (27.5 million square kilometers) in area.

Can the **ozone hole be healed**?

Yes. While the latest figures represent an increase in the ozone hole's size over previous years, there is some good news: compared to the 1980s, the hole is expanding more slowly. If we continue to reduce pollutants, the expansion may eventually stop and be reversed. Scientists believe that, if this happens, it will take about 50 years for ozone levels to return to natural levels.

WIND

What is **wind**?

Wind, simply put, is the movement of air in the atmosphere. Wind movement is caused by the fact that air will move from high pressure areas to low pressure areas.

When a weatherman says that the wind direction is "westerly" does that mean the wind is coming from the West or that it is blowing towards the West?

Wind directions expressed by meteorologists and the National Weather Service indicate which direction the wind is coming *from* rather than where it is blowing *to*. So, if an area is experiencing "northwest winds," for example, it means that the wind is coming from the northwest and blowing toward the southeast.

In other words, high pressure zones contain more densely packed molecules of various gases, which tend to flow to areas where the air is less dense. This concept was first explained by the Greek philosopher Anaximander (c. 610 B.C.E.–c. 546 B.C.E.), who explained that wind was a natural phenomenon and not caused by the gods or by trees waving their leaves, as some people thought.

Does **wind** have a lot of **energy** that could help reduce the need for fossil fuels?

If people could harvest all the energy in the Earth's wind through the use of windmills, for instance, it would generate about 3.6 million kilowatts of power, enough to supply the energy needs of 3.6 billion Americans. Since Americans use much more energy than most people on the planet, it is safe to say that the energy needs of the nearly seven billion people on Earth could be met by wind power alone. Unfortunately, it would be impossible to extract *all* of this energy. Wind turbines have become economically feasible, but we could never place them over every land and sea surface on the planet.

What does it mean when we say something is on the **lee side** of the wind?

If a person is standing on the lee side of something (say, a building or rocky prominence), then he or she is protected from the wind because that obstacle is between him or her and the oncoming wind.

What are **lee troughs** and **lee depressions**?

A lee trough—also known as a dynamic trough—is a low-pressure zone that forms downwind from a north-south mountain range. A lee depression is essentially the same thing, except that troughs are long and stretched out, while depressions are well-defined, localized areas of low pressure.

What is **anemophobia**?

A fear of the wind—and sometimes even mere drafts—is known as anemophobia.

Climatologist William Ferrel has been called "The Father of Geophysical Fluid Dynamics." (*NOAA*).

What is **Buys Ballot's Law**?

Dutch meteorologist and chemist Christoph Hendrik Diederik Buys Ballot (1817–1890) was a pioneer in meteorology, especially when it came to explaining how air flows in large weather systems. The law that bears his name refers to the fact that when you stand with your back to the wind, the air pressure will be lower on your left than on your right in the Northern Hemisphere, and the opposite is true if you are standing in the Southern Hemisphere. This phenomenon—which actually only proves to be true during well-organized weather systems—was also discovered by American climatologist William Ferrel (1817–1891). Ferrel actually formulated this theory a few months before Buys Ballot. The Dutchman graciously acknowledged that the credit deserved to go to Ferrel, but the "law" had already been denoted Buys Ballot's Law, and the name stuck.

What are **trade winds**?

Trade winds are very consistent winds blowing through the tropics (between 30° south and 30° north latitudes) at about 11 to 14 miles (18 to 22 kilometers) per hour, sometimes for days on end. In the Northern Hemisphere they blow toward the equator from the northeast, and south of the equator they blow in from the southeast. While the term "trade winds" leads most people to think they got their names from the days when large sailing ships depended on them for shipping routes, the word "trade" actually has a German origin and means "track" or "path."

When were the phenomenon of **trade winds** first explained, and by whom?

Astronomer Edmund Halley (1656–1742), who is usually thought of as the discoverer of the comet that bears his name, was also interested in cartography, oceanography, and the atmosphere. For instance, he created tidal charts and maps illustrating the path of ecliptic shadows. In 1868, he formed a theory to explain why we have the trade winds. Halley correctly guessed that it had to do with warm tropical air mixing with cooler air from more northern and southern latitudes. His idea, though, did not adequately explain why the winds blow from east to west, rather than south to north, as his theory would have indicated. It took English meteorologist George Hadley's (1685–1768) discovery of convection cells to amend the theory correctly in 1735.

What are aeolian sounds?

The somewhat musical, sometimes mournful sound wind makes when blowing across branches, wires, or circular objects is called an aeolian sound. Aeolus was the Greek god of wind, and aeolian is a musical term used to refer to wind instruments or to a diatonic scale.

What are the **westerlies**?

These winds flow at mid-latitudes (30 to 60 degrees north and south of the equator) from west to east around the Earth. The high-altitude winds known as the jet stream are also westerlies.

What is a **Chinook**?

A Chinook, sometimes called a "snow eater" because it melts snows, is a wind that is generally warm and originates from the eastern slope of the Rocky Mountains. it often moves from the southwest in a downslope manner, causing a noticeable rise in temperature that helps to warm the plains just east of the Rocky Mountains.

What is **"The Doctor"**?

This is an affectionate term, used in places such as the Caribbean, referring to the cool, refreshing sea breezes that make a hot day more bearable.

What is a **harmattan**?

This word refers to the sub-Saharan, west coast winds that are dry, hot, and dusty, but fairly moderate in strength.

What is a **katabatic wind**?

A katabatic wind develops because of cold, heavy air spilling down sloping terrain (e.g., a Chinook), moving the lighter, warmer air in front of it. The air is dried and heated as it streams down the slope. At times, the falling air becomes warmer than the air it restores.

What are some **other katabatic winds** called?

In addition to the Chinook, the Santa Ana winds that flow down the Sierras in southern California, and the Taku, which is a frigid wind in Alaska, are both examples of katabatic winds.

What is a **Nor'Easter**?

A Nor'Easter is a storm along the eastern coast of North America that affects the region with northeasterly winds with speeds up to 75 miles (121 kilometers) per hour or more. Such storms evolve when low pressure systems accumulate humid air from the Atlantic Ocean, or from the Gulf of Mexico, and combine it with cold dry air coming down from Canada in conjunction with a strong jet stream. The system rotates counterclockwise, bringing strong rain storms in the south and, in winter, snow to the Northeast.

Nor'Easters have wreaked havoc on the United States a number of times. For example, a February 1969 Nor'Easter dumped 70 inches (178 centimeters) of snow on Rumford, Maine, and 164 inches (416.5 centimeters) on Pinkham Notch, New Hampshire. One of the worst Nor'easters resulted in a super storm in March 1993 that is still referred to as the "Storm of the Century."

What are **Northers**?

Also known as Blue Northers, these are cold winds that blow southward from the Texas plains down to the Gulf of Mexico.

What are the **Santa Ana winds**?

When a high pressure system forms over the Great Basin—usually during the fall—it pushes air masses downwards, causing compression that results in winds in California that elevate temperatures upwards of 100°F (38°C). These winds can reach speeds well over 75 miles (120 kilometers) per hour in many cases, blowing down into the densely populated coastal regions. The dry winds of the Santa Ana help create conditions that make brush fires in the hills around Los Angeles and other cities much more likely. In fact, one of the most disastrous fires in California's history—known as the Cedar Fire—occurred in October 2003 as a result of strong Santa Ana winds. The fire destroyed 721,791 acres (2,921 square kilometers) of land and burned 3,640 homes to the ground.

What are some other **named winds** and where are they located?

There are many regions of the world that experience regular winds that have distinctive characteristics, such as wind direction, temperature, and humidity levels. Including some of those already mentioned above, here is a list of the popular names people give to their local winds.

Common Names for Local Winds

Wind Name	Location	Winds From
Chinook	Western U.S.	West
Santa Ana	Southeastern U.S.	Northeast
Norther	Central U.S.	North
Papagayo	Mexico	Northwest
Norte	Mexico	North
Terral	Western South America	Northeast
Virazon	Western South America	Southwest

Which U.S. state benefits from very favorable wind conditions?

The state of Hawaii enjoys a climate where ocean breezes naturally cool the islands, which would otherwise be quite hot because of the state's tropical location. Because of these ocean breezes, temperatures rarely exceed 90°F (32°C), and the evenings are relatively cool, even during the summer.

Wind Name	Location	Winds From
Zonda	Southeast South America	Northwest
Pampero	Southern South America	Southwest
Mistral	France	North
Bise	Germany	North
Bora	Eastern Europe	Northeast
Tramontana	Eastern Europe	North
Levanter	Italy, Spain	East
Leveche	Western Sahara, Spain	South
Chili	Western Sahara	Southwest
Gibli	Western Sahara	Southwest
Leste	Western Sahara	Southeast
Harmattan	Western Sahara	Northeast
Sirocco	Central Sahara	South
Khamsin	Egypt, Sudan	South
Haboob	Sudan	Southwest
Berg	South Africa	Northeast
Shamal	Middle East	Northwest
Seistan	Iran, Turkmenistan, Uzbekistan	North
Nor'wester	India, Pakistan, Afghanistan	Northwest
Karaburan	Western China	Northeast
Buran	Siberia, Central Russia	Northeast
Purga	Siberia, Eastern Russia	Northeast
Bohorok	Malaysia, Indonesia	West
Koembang	Indonesia	West
Brickfielder	Southeastern Australia	North
Southerly Buster	Southeastern Australia	Southwest

What were the **fastest winds** ever recorded?

During a tornado strike in Moore, Oklahoma, on May 3, 1999, a wind speed of 318 miles (512 kilometers) per hour was estimated by video recordings showing debris movement. Other incredible winds include a measurement of 268 miles (431 kilometers) per hour during a tornado incident in Red Rock, Oklahoma, on April 26, 1991. Not all wind records require tornadoes, though. Mount Washington in New

The effects of the jet stream are clearly seen in this photo showing a mass of moist air being carried by high winds near the North American West Coast. (*NOAA*)

Hampshire is considered one of the windiest places on Earth. Winds frequently blow there at speeds exceeding 100 miles (160 kilometers) per hour, and one gust was recorded at 231 miles (372 kilometers) per hour.

What is a **mauka breeze**?

A mauka breeze is the Hawaiian term for cooling winds that sweep down from the volcanic mountains to cool the warmer, lower regions of the islands.

What is the **jet stream**?

The jet stream is a band of swiftly moving air located high in the atmosphere and affects the movement of storms and air masses closer to the ground. The currents of air flow from west to east and are usually a few miles deep, up to 100 miles (160 kilometers) wide, and well over 1,000 miles (1,600 kilometers) in length. The speed at which the air travels in the jet stream is over 57.5 miles (92 kilometers) per hour, sometimes moving as fast as 230 miles (386 kilometers) per hour.

There are two polar jet streams, one in each hemisphere. The jet streams meander across the troposphere and stratosphere (up to 30 miles [48 kilometers] high) and between 30 and 70 degrees latitude. There are also two subtropical jet streams (one in each hemisphere) that range between 20 and 50 degrees latitude. The subtropical streams flow between altitudes of 30,000 to 45,000 feet (9,150 to 13,700 meters) and are even swifter than the polar streams, moving at speeds of over 345 miles (550 kilometers) per hour.

What is a **Rossby wave**?

Named after Swedish meteorologist Carl-Gustaf Rossby (1898–1957), Rossby waves refer to the large air mass waves that inhabit the middle layer of our atmosphere, including the jet stream and high- and low-pressure systems.

How was the **jet stream discovered**?

With the advent of airplanes that could cruise at elevations of over 30,000 feet (over 9,000 meters) high, pilots—such as World War II bomber pilots flying over Japan and the Mediterranean Sea—discovered the effects of the jet stream on their aircraft.

What is the **low-level jet stream**?

Seen in the central United States, low-level jet streams are air flows coming in from the Gulf of Mexico into the Central Plains, but they also occur as streams flowing from the Indian Ocean into Africa. Low-level jet streams occur at altitudes of only a couple thousand feet and can bring in moisture and warm air that creates severe thunderstorms and tornadoes. In the Central Plains, though, they only occur at night.

What is the **arctic oscillation**?

Arctic oscillation is a measure of the differences in air pressure between air within the Arctic Circle and air that is between the circle and about 55°N longitude. The arctic oscillation can be either positive (air pressure is lower over the arctic region) or negative (the air pressure is higher). In the case of the former, winds become stronger across mid-latitude regions, Eurasia becomes warmer, and drier conditions prevail in the American West and the Mediterranean. Also, storms move farther north into Alaska and northern Europe. When the arctic oscillation is negative, the opposite is mainly true, and the American West Coast and the Mediterranean experience wet weather, while Eurasia cools.

What is **wind shear**?

Wind shear refers to rapid changes in wind speed and/or direction over short distances and is usually associated with thunderstorms. Sometimes, though, wind shears can also occur as a result of a strong front moving through a region, or from abrupt changes in air mass near mountains. Wind shear is especially dangerous to aircraft, and Doppler radars at airports help to warn pilots of this threat.

What is the effect of a **microburst** on **aircraft**?

Microbursts are downbursts of air with a diameter of 2.5 miles (4 kilometers) or less. Often associated with thunderstorms, they can generate winds of hurricane force that change direction abruptly. Headwinds can become tailwinds in a matter of seconds, forcing aircraft to lose air speed and altitude. After microbursts caused several major air catastrophes in the 1970s and 1980s, the Federal Aviation Administration (FAA) installed warning and radar systems at airports to alert pilots when

Air particles clearly reveal a dangerous microburst. Sometimes, airplanes can be caught in these dangerous bursts of air. (*NOAA Photo Library, NOAA Central Library; OAR/ERL/National Severe Storms Laboratory*)

conditions were right for wind shears and microbursts.

What **causes turbulence**?

Air turbulence usually occurs in the higher levels of the atmosphere, which is why you don't notice it unless you're in an airplane. It happens when upward- and downward-moving currents of air mix (convective mixing). This typically is noticed while flying through a cloud or near a jet stream.

Why do meteorologists refer to the **Reynolds number** when talking about turbulence?

The Reynolds number is a mathematical result computed by calculating the ratio of inertial to viscous forces. Put in more understandable English, it measures how fluids move through an area of defined diameter. It is named after English physicist and engineer Osborne Reynolds (1842–1912), who was interested in how water flows through rivers and in waves and tides. However, it can also be used in terms of air flowing through the atmosphere, and thus has applications in meteorology, where Reynolds's formula is used to calculate air turbulence.

What is an **"air pocket"**?

Air pockets are caused by turbulent air. Many people who have ridden in airplanes are familiar with them, as they cause that bouncing sensation as you hit an updraft, followed abruptly by a downdraft, in quick succession.

How does **air flow** around **low and high pressure systems**?

Air tends to flow toward a low pressure system and away from a high pressure system. In the Northern Hemisphere, the air will spiral in a counterclockwise direction as it moves toward a low pressure center, and it will move in a clockwise direction as it shifts away from a high pressure center.

What is the **Intertropical Convergence Zone**?

The Intertropical Convergence Zone (ITCZ) is a band around our planet where the winds from the Southern and Northern Hemispheres rub against each other. The

What are the "Doldrums"?

Sailors used to call the low-pressure area near the equator the "Doldrums" because winds here tend to be very light and variable and, thus, difficult to sail in. The Doldrums are found near where the trade winds originate. The expression "in the doldrums" has come to mean a person who is feeling inactive, listless, or lethargic.

resulting convergence causes air to rise and is an ideal place for tropical storms to form. Generally near the equator, the ITCZ changes position depending on the effects of the Sun and seasonal cycles. Usually, there is one ITCZ, but sometimes a double ITCZ can form. Land beneath the ITCZ receives much more rain (up to 200 days annually) than land areas outside the zone, and shifts in the zone can lead to wet or dry weather.

What is the **moist tongue**?

Sometimes meteorologists come up with some pretty fanciful descriptive terms. Moist tongue falls into that category! It refers to humid tropical air moving in the direction of either the North or South Pole. A moist tongue regularly occurs in the spring and summer over the U.S. Central Plains, where it works in conjunction with a jet stream to produce storms.

What are the **horse latitudes**?

The horse latitudes are two high pressure belts characterized by low winds about 30 degrees north and south of the equator. Dreaded by early sailors, these areas have undependable winds with periods of calm. In the Northern Hemisphere, particularly near Bermuda, sailing ships carrying horses from Spain to the New World were often becalmed. When water supplies ran low, these animals were the first to be rationed water. Dying from thirst, the animals were tossed overboard to conserve water for the men. Explorers and sailors reported that the seas were "strewn with bodies of horses," which may be why the areas are called the horse latitudes. The term might also be rooted in complaints by sailors who were paid in advance and received no overtime when the ships slowly traversed the area. During this time, they were said to be "working off a dead horse."

WIND STORMS

Who **developed** the **Beaufort Scale**?

Originally created in 1806 by Irish-born hygrographer and Rear Admiral Sir Francis Beaufort (1774–1857), the Beaufort Scale is a subjective way to measure wind speeds. Although the anemometer—a device for measuring wind speeds—had

Beaufort Scale

Beaufort Number	Description	Wind Speed mph/kph/knots	Wave Height Feet/Meters	Land/Sea Description
0	Calm	<1/<1/<1	0/0	Calm; smoke rises vertically / Flat seas
1	Light air	1-3/1-5/1-2	0.33/0.1	Wind motion visible in smoke / Ripples without crests
2	Light breeze	3-7/6-11/3-6	0.66/0.2	Wind felt on exposed skin; leaves rustle / Small wavelets; crests of glassy appearance, but don't break
3	Gentle breeze	8-12/12-19/7-10	2/0.6	Leaves and smaller twigs in constant motion / Large wavelets; crests begin to break; scattered whitecaps
4	moderate breeze	13-17/20-28/11-15	3.3/1	Dust and loose paper raised; small branches begin to move / Small waves
5	Fresh breeze	18-24/29-38/16-20	6.6/2	Moderate-size branches move; small trees sway / Moderate, longer waves (about 4 feet/1.2 meters); some foam and spray
6	Strong breeze	25-30/39-49/21-26	9.9/3	Large branches in motion; whistling heard in overhead wires; umbrella use becomes difficult; empty plastic garbage cans tip over / Large waves with foam and crests and some spray
7	High wind, Moderate Gale, Near Gale	31-38/50-61/27-33	13.1/4	Whole trees in motion; effort needed to walk against the wind; swaying of skyscrapers may be felt, especially by people on upper floors / Sea heaps up and foam begins to be blown in streaks in wind direction
8	Fresh Gale	39-46/62-74/34-40	18/5.5	Twigs break off trees; drivers feel their cars being noticeably blown by winds / Moderately high waves with breaking crests, forming spindrift; streaks from foam
9	Strong Gale	47-54/75-88/41-47	23/7	Larger branches break off trees and some small trees blow over; construction and temporary signs and barricades blow over; damage to circus tents and canopies / High waves with dense foam; wave crests start to roll over; considerable spray

Beaufort Number	Description	Wind Speed mph/kph/knots	Wave Height Feet/Meters	Land/Sea Description
10	Whole Gale/Storm	55-63/89-102/48-55	29.5/9	Trees are broken off or uprooted, saplings bent and deformed, poorly attached asphalt shingles and shingles in poor condition peel off roofs / Very high waves; large patches of foam from wave crests give the sea a white appearance; considerable tumbling of waves with heavy impact; large amounts of airborne spray reduce visibility
11	Violent Storm	64-75/103-117/56-63	37.7/11.5	Widespread vegetation damage; more damage to most roofing surfaces, asphalt tiles that have curled up and/or fractured due to age may break away completely / Exceptionally high waves; very large patches of foam, driven before the wind, cover much of the sea surface; very large amounts of airborne spray severely reduce visibility
12	Hurricane-force	$\geqslant$76/$\geqslant$118/$\geqslant$64	$\geqslant$46/$\geqslant$14	Considerable and widespread damage to vegetation, some windows broken, structural damage to mobile homes and poorly constructed sheds and barns; debris may be hurled about / Huge waves; sea is completely white with foam and spray; air is filled with driving spray, greatly reducing visibility

In April 2001 a huge dust storm that began in China reportedly blocked out sunlight in Jilin Province. It sent plumes of dust high into the atmosphere that were eventually carried as far as the Great Lakes. This image was taken by NASA's Terra satellite. A 2005 study showed that large dust storms are now nearly annual events in China, though historically they used to occur only every 30 years. (*NASA*)

already been invented by Robert Hooke (1635–1703), in Beaufort's day the device was still not widely used. Beaufort joined the navy in 1790, and by 1805 had risen to command his own ship. He had picked up hygrography along the way, and while off the coast of South America, commanding the H.M.S. *Woolwich,* developed his "Wind Force Scale and Weather Notation," while conducting surveys. This scale did not use any new terminology, but rather represented Beaufort's attempt to create standardized notation that could be used consistently by scientists. While highly praised by his superiors for his extremely detailed records, Beaufort, who became the Admiralty's official hygrographer in 1829, was not successful in getting the Royal Navy to adapt his standards until 1838. He was knighted in 1848 and retired in 1855. Beaufort's scale was modified in 1874 by the International Meteorological Committee, which added details about wind effects on both land and sea.

What are the hazards of **dust storms**?

Dust storms regularly occur in such places as the Sahara and Gobi Deserts. Not only do dust storms destroy farms, they also invade homes and other buildings and can cause health problems ranging from stinging eyes to lung and other respiratory problems. Lowering visibility, dust storms have been known, too, to cause traffic accidents just as well as a snow storm can. In 1995, for example, eight people were killed on the New Mexico–Arizona border when a dust storm blew across a highway.

How **far** can a **dust storm travel**?

Central Asia and China have annual dust storms in their desert areas almost every April, and the dust from these storms has been known to reach as far away as Hawaii. On the other hand, dust storms from Africa have blown dust all the way to Florida. Blowing dust can actually be a good thing, as it helps distribute soil nutrients around the planet. For example, it is known that rainforest soil in the Amazon is replenished from African soil, as is soil in the American southeast.

Where do **haboobs** occur?

Derived from the Arabic word *habb,* meaning "to blow," a haboob is a violent dust storm. Haboobs are common in the Sahara Desert, as well as in the deserts of Australia, Asia, and North America.

What is the **windiest place** on Earth?

Mt. Washington, New Hampshire, is the windiest place on Earth that has a weather station to record such data. In 1934, wind speeds were clocked at 231 miles (372 kilometers) per hour; the average annual wind speed there is 35.3 miles (56.8 kilometers) per hour. While no official measurements are available, the windiest places in the world are probably found in the coastal regions of Antarctica.

Is **Chicago** really the **"windy city?"**

Chicago is not the windiest city in the United States. It actually got the nickname of "The Windy City" because of its reputation for being the home of blowhard politicians. Chicago's average wind speed of 10.4 miles (16.7 kilometers) per hour is exceeded by Boston (12.5 [20.1]), Honolulu (11.3 [18.2]), and Dallas and Kansas City (both 10.7 [17.2]).

HEAT AND COLD

MEASURING TEMPERATURE

How was the **Fahrenheit scale** developed?

The temperature scale still in use in the United States (while the rest of the world uses the metric system of Celsius), is named after German engineer and physicist Gabriel Fahrenheit (1686–1736). Fahrenheit developed his scale after visiting Danish astronomer Ole Christensen Rømer (1644–1710) in 1708. For his experiments, Rømer was using an alcohol thermometer upon which he had marked a scale from zero degrees (the lowest temperature he could achieve in his laboratory using a mixture of ice, water, and salt) to 60 degrees, the boiling point of water. This system actually mimics the basic idea behind the Celsius scale.

While Fahrenheit left no records as to why he chose certain high and low points on his thermometers, there has been speculation on this matter. It is known that he used his own body temperature as the high point on his thermometer, but he marked this temperature as 96 degrees. Some believe he chose this number because it was easily divisible by twos and threes. For his zero degree mark he used Rømer's figure. Fahrenheit later determined that water normally froze at 32 degrees on his scale and boiled at 212 degrees.

What is the **Kelvin** temperature scale?

The Kelvin scale is typically used in laboratories dealing with extremely cold temperatures. Zero degrees Kelvin indicates absolute zero, which is the point at which molecular motion stops. The scale is named after the 1st Baron Kelvin, William Thomson (1824–1907), who was a British engineer and physicist. Thomson developed the absolute thermometric scale, which used degrees Celsius as its increments and had absolute zero being –273°C (the actual figure has now been deter-

Thermometers in the United States display both Celsius and Fahrenheit temperatures, which saves people the trouble of having to convert degrees.

mined at –273.15°C. Such extremes are rarely needed in meteorological studies, however.

Why was the name of the **Centigrade scale** changed to **Celsius**?

The centigrade temperature scale was created by Swedish astronomer Anders Celsius (1701–1744) in 1742. He was determined to create a new temperature scale as an international standard to be used by scientists. To do so, he wanted the degrees of zero and 100 to be set under conditions easily reproduced in the laboratory: and those were the freezing and boiling points of water. Because the two extremes were divided into 100 degrees, he named it the centigrade scale. To honor his accomplishment, the Ninth General Conference on Weights and Measures officially changed the name to Celsius in 1948. Fortunately, since both words begin with the same letter, °C did not have to be changed as well.

How do I **convert Fahrenheit** to **Celsius** or **Kelvin**?

Fahrenheit and Celsius are two common temperature scales used throughout the world. Temperature in Fahrenheit can be converted to Celsius by subtracting 32 and multiplying by five; divide that number by nine and you have Celsius. Conversely, you can convert Celsius to Fahrenheit by adding 32, multiplying by nine and finally dividing by five. Kelvin, a system used by scientists, is based on the same scale as Celsius. All you have to do is add 273 to your Celsius temperature to obtain Kelvin. Zero degrees Kelvin is negative 273°C.

What is a **low high** temperature and a **high low** temperature?

When meteorologists look at daily temperature, there is always a low and a high temperature for each day. If the high temperature is the coldest high temperature for that day or for the month on record, you have a new record—a new low high. Conversely, if the low temperature for a day is quite warm and breaks records, that's a new high low!

What **chemicals** are used in **thermometers**?

The two liquids that are typically used in a thermometer are mercury or alcohol. The alcohol in thermometers is dyed red to be easier to read, while mercury has a silver

Can crickets be used as thermometers?

It has long been known that crickets chirp at different rates depending upon the temperature. While calculating temperatures based on this is not ideal, it can serve in a pinch if you find yourself without a thermometer—and provided it's not the dead of winter. The general rule is to count how many times a cricket chirps during a 14-second interval. Add 40 to that figure, and the result is degrees in Fahrenheit. If you have more patience, count the number of chirps over the period of one minute. Take the number of chirps you hear, subtract 72, divide by 4, and add the result to 60, which gives you the degrees in Fahrenheit ([chirps – 72] / 4 + 60 = degrees F). For example, if you hear 68 chirps, then 68 minus 72 equals –4, divided by 4, equals –1. Add this to 60 and the result is 59°F.

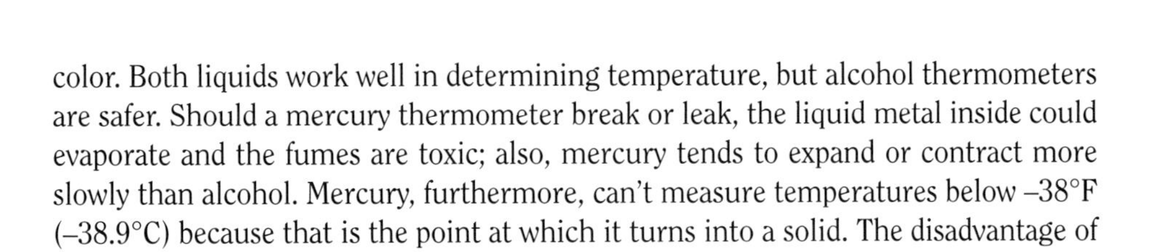

color. Both liquids work well in determining temperature, but alcohol thermometers are safer. Should a mercury thermometer break or leak, the liquid metal inside could evaporate and the fumes are toxic; also, mercury tends to expand or contract more slowly than alcohol. Mercury, furthermore, can't measure temperatures below –38°F (–38.9°C) because that is the point at which it turns into a solid. The disadvantage of alcohol, however, is that it tends to evaporate or polymerize over time.

What are **thermographs** and **hygrothermographs**?

A thermograph is a device that records changes in temperature over time by drawing a line over a rotating chart in a manner similar to a seismograph. A hygrothermograph does the same thing, except it also indicates levels of humidity.

Does **Alaska get hotter** than **Hawaii**?

Strangely enough, the two states are currently tied. Hawaii, though located in the tropics, enjoys very moderate temperatures due to favorable winds and the islands' topography. The mountainous parts of this fiftieth state can actually get very cold, and the tops of Mauna Kea and Mauna Loa receive snowfall regularly. While the average temperatures across the state of Hawaii are higher than in Alaska, the islands have fewer extremes. Alaska's interior region can get very warm indeed, sometimes reaching 90°F (32°C). The highest temperature in Alaska thus far recorded is 100°F (37.8°C), which was measured at Ft. Yukon on June 27, 1915. On April 27, 1931, Pahala, Hawaii reached the same extreme.

Does the **North Pole** have the same **average temperatures** as the **South Pole**?

One might think that the northern and southern extremes of the planet would have comparable annual temperatures. In actuality, the South Pole is much colder. This

What is a swamp cooler?

A swamp cooler is an alternative to a traditional air conditioner. A traditional air conditioner uses compressed Freon gas to cool the air. This system works very well in most climates where the humidity is above 30 percent. In drier climates (below 30 percent humidity) a swamp cooler is very effective. Also called an evaporative cooler, a swamp cooler works sort of like a wet bulb thermometer. Pads inside the cooler absorb water, and a fan then blows on the pads, cooling the air and making it more humid. Swamp coolers are more energy efficient that regular air conditioners because the only electricity needed is what is used for the fan; no compressor is required.

is because giant glaciers have built up on the continent of Antarctica, where the South Pole is located. The inland ice sheets of Antarctica are 8,036 feet (2,450 meters) thick on average, and this thick layer of ice keeps the South Pole about 50°F (23.6°C) colder than in the North.

What is a **diurnal temperature range**?

As one might guess from the term, diurnal temperature range means the variation between the lowest and highest temperatures during a given day at a certain location. This can vary widely, depending on the season and place. For example, the climate on the Front Range of the Rocky Mountains in North America has been known to change by as much as 100 degrees within a 24-hour period, while in the tropics temperatures tend to remain more stable, changing by, perhaps, 10 degrees or so within a specific day.

What is a **temperature inversion**?

Simply put, this is when temperatures in the troposphere (lowest layer of the atmosphere) are colder at lower altitudes than at higher altitudes, which is just the opposite of what normally occurs. Inversions can occur as the result of weather fronts moving through an area, or because of winds blowing over ground that is frozen over with ice or snow, or because of frigid lake or ocean conditions. When an inversion occurs it prevents air from circulating vertically, which can result in pollutants being trapped at low altitudes. Inversions also disrupt radio and radar transmissions.

What are the **wet bulb** and **dry bulb temperatures**?

Taking a reading of the wet bulb temperature is a way of measuring humidity. It is the temperature of a volume of air that has been cooled adiabatically until the humidity is at 100 percent and then compressed to the original pressure of the system prior to cooling. Using a wet bulb thermometer provides one reading, while a dry bulb thermometer is used for the second reading. The greater the difference between the temperatures read on the two thermometers, the less humid the air is. Using the two in conjunction is a single device called a psychrometer. The psychrometer measures the difference between the wet and dry bulb, and then, using established tables, the humidity of the air can be calculated.

How is a **wet bulb thermometer constructed**?

The basic design of a wet bulb thermometer is to insert a thermometer into a reservoir of pure, distilled water. Wrapped around the thermometer is a piece of cloth (usually muslin) that acts as a wick, drawing water up through the reservoir via capillary action. As the moisture in the cloth above the surface of the water evaporates, it draws heat from the thermometer, thus lowering the temperature until the surrounding air reaches a point of saturation. A dry bulb thermometer is simply a regular thermometer that is not dipped into the water reservoir or wrapped in cloth.

What **types of psychrometers** are there?

Old-style sling psychrometers are made from two thermometers secured to metal that is allowed to hang for a number of minutes until the wet bulb begins to cool. More modern psychrometers use a variety of sensors, including the use of chemicals that change their electrical conductivity due to humidity levels.

What are **cooling degree days** and **heating degree days** and what do they have to do with **air conditioning**?

Utility companies use a measurement called "cooling degree days," which estimates how many days people are likely to run their air conditioners yearly.

Utility companies in the United States use the term "cooling degree days" to refer to the number of days when it is likely that air conditioners will be turned on, and "heating degree days" to determine when furnaces are likely to run. Because it would be impractical to visit every home and business to see whether the air conditioner or furnace was running, utility companies assume that customers are most comfortable when the temperature is 65°F (18.3°C) and that they will run their air systems accordingly. A "degree day" is not actually a 24-hour period. Rather, it is a measurement of the difference between the optimum of 65°F (Fahrenheit is used as the standard because it is a U.S. system)

and the mean (average) temperature of a particular day. So, for example, if the mean temperature in Dallas, Texas, on a warm July day is 85°F, the utility companies will count that as 20 cooling degree days (85°F – 65°F = 20 cooling days). The states with the most annual cooling days on average are Florida, Texas, Arizona, California, and other southern states that, on average, rack up about 4,000 cooling days annually. Heating degree days use the same concept to calculate furnace use. As one might imagine, northern states log more heating degree days than southern states.

What **month** is known for the most **extreme changes in temperature**?

July holds the honor for most extreme variations of temperature globally. Record extremes on the planet occurred when measuring the temperature in Libya versus the temperature at the South Pole. The difference between the two was 265°F (129.4°C).

What is the **"120 Club"**?

Any U.S. state that has recorded a temperature at or above 120°F (48.9°C) is a member of the "120 Club." Nine states have met this dubious requirement:

- California: 134°F/56.7°C
- Arizona: 128°F/53.3°C
- Nevada: 125°F/51.7°C
- New Mexico: 122°F/50°C
- Kansas: 121°F/49.4°C
- North Dakota: 121°F/49.4°C
- Oklahoma: 120°F/48.9°C
- Arkansas: 120°F/48.9°C
- South Dakota: 120°F/48.9°C
- Texas: 120°F/48.9°C

What is the **"60 Below Club"**?

Given the above answer, this is pretty easy to guess. States that have recorded temperatures of –60°F (–51°C) or below belong to this chillin' club.

- Alaska: –80°F/–62.2°C
- Montana: –70°F/–56.6°C
- Utah: –69°F/–56°C
- Wyoming: –63°F/–54.4°C
- Colorado: –61°F/–51.6°C
- Idaho: –60°F/–51°C
- Minnesota: –60°F/–51°C
- North Dakota: –60°F/–51°C

HEAT

What is a **heat wave**?

A heat wave is a period of two or more days in a row when apparent temperatures on the National Weather Service heat index exceed 105°F to 110°F (40°C to 43°C). The temperature standards vary greatly for different locales. Heat waves can be extremely dangerous. According to the National Weather Service, 175 to 200 Americans die from heat in an average summer. Between 1936 and 1975, as many as 15,000 Americans died from problems related to heat. In 1980, 1,250 people died during a brutal heat wave in the Midwest. in 1995, more than 700 people died in the city of Chicago from heat-related problems. A majority of these individuals were the elderly living in high-rise apartment buildings without proper air conditioning. Large concentrations of buildings, parking lots, and roads create an "urban heat island" in cities.

What is the **heat index**?

The heat index is a measurement of what hot weather feels like to the average person for various temperatures and relative humidities. Heat exhaustion and sunstroke are inclined to happen when the heat index reaches 105°F (40°C). The reason why it feels hotter to people when it is more humid is that the body cools itself by perspiring. Beads of perspiration help cool the skin as they evaporate, but when the air is too humid to allow for this evaporation, this natural cooling system no longer works well. The danger point for humans can occur when the heat index is between 90°F (32°C) and 104°F (40°C), and when the heat index reaches above these temperatures, weather conditions become very dangerous indeed. At this extreme, heat exhaustion, heat stroke, or sunstroke may occur quickly, especially during physical activity, and it is very important to stay in the shade and remain hydrated. The chart below provides the heat index for some temperatures and relative humidities.

If you really want to fry an egg on the ground, try doing so on asphalt on a hot summer's day.

How **hot** can it get on the **Earth's surface**?

Temperature records used by meteorologists and reported in weather broadcasts refer to air temperature. The surface of the Earth, however, can get much hotter than the record 120s and

Heat Index (°F/°C)

Relative Humidity	70/21	75/23.9	80/26.7	85/29.4	90/32.2	95/35	100/37.8	105/40.5	110/43.3	115/46.1	120/48.9
	Actual Air Temperature										
	Feels Like										
0%	64/17.8	69/20.5	73/22.8	78/25.5	83/28.3	87/30.5	91/32.8	95/35	99/37.2	103/39.4	107/41.7
10%	65/18.3	70/21.1	75/23.9	80/26.7	85/29.4	90/32.2	95/35	100/37.8	105/40.5	111/43.9	116/46.7
20%	66/18.9	72/22.2	77/25	82/27.8	87/30.5	93/33.9	99/37.2	105/40.5	112/44.4	120/48.9	130/54.4
30%	67/19.4	73/22.8	78/25.5	84/28.9	90/32.2	96/35.5	104/40	113/45	123/50.5	135/57.2	148/64.4
40%	68/20	74/23.3	79/26.1	86/30	93/33.9	101/38.3	110/43.3	123/50.5	137/58.3	151/66.1	
50%	69/20.5	75/23.9	81/27.2	88/31.1	96/35.5	107/41.7	120/48.9	135/57.2	150/65.5		
60%	70/21.1	76/24.4	82/27.8	90/32.2	100/37.8	114/45.5	132/55.5	149/65			
70%	70/21.1	77/25	85/29.4	93/33.9	106/41.1	124/51.1	144/62.2				
80%	71/21.7	78/25.5	86/30	97/36.1	113/45	136/57.8					
90%	71/21.7	79/26.1	88/31.1	102/38.9	122/50						
100%	72/22.2	80/26.7	91/32.8	108/42.2							

Is it possible to fry an egg on the sidewalk?

Well, not really. Concrete sidewalks are not actually the best place to cook an egg (not to mention they aren't very sanitary). On very hot days, the sidewalk temperature can get to about 145°F (62.8°C), but an egg requires about 158°F (70°C) to cook thoroughly. On the other hand, if you put an egg on blacktop, you are more likely to be successful because the black surface will absorb more heat. Even better is using a metal surface, and people have been known to cook an egg on the hood of a car on a summer's day.

130s°F (high 40s to high 50s°C) that are usually considered the extreme. Soil temperatures can actually rise about 180°F (82°C), and the surface temperature in Death Valley has been measured at a hellish 200°F (93°C)

How **dangerous** are **heat waves** to people?

Heat waves are responsible for some of the most lethal weather-disasters in history. In fact, in the United States the heat wave of 1980 caused more deaths (between 10,000 and 15,000, depending on sources) than the 1900 hurricane that killed between 8,000 and 12,000 people in Galveston, Texas. More recently, in 2003 a heat wave in Europe struck, killing somewhere between 35,000 and 50,000 people. The heat had a particularly lethal effect on the people of France, where over 14,800 people died of heat-related problems, mostly heat stroke. In France, fewer people have air conditioning in their homes, and many elderly living in apartments succumbed to the suffocating temperatures.

What are some of the **worst summers** in terms of **heat-related deaths** that have occurred in the **United States?**

The United States has also endured some scorching summers that resulted in thousands of deaths. Below are some of the worst on record.

Deadly Summers in the United States

Year	Number of Deaths
1901	9,508
1936	4,678
1975	1,500-2,000
1980	1,700
1988	5,000-10,000

What is the **hottest place** in the **United States**?

The most inhospitably hot and dry place in America is Death Valley in eastern California. Also the lowest place, at 280 feet below sea level, Death Valley regularly

Death Valley, California, is the hottest place in the United States, frequently seeing temperatures over 100°F (38°C).

experiences temperatures in the 120s Fahrenheit (high 40s to low 50s Celsius) and higher, and the valley sees temperatures above 100°F (37.8°C) between 140 and 160 days every year, even in the shade. The record high here has been tallied at 134°F (56.7°C). Annual rainfall in Death Valley is less than two inches (50 millimeters).

What are some of the most **unusually hot temperatures** experienced in the **United States**?

People sometimes fret about global warming when they experience an unusually warm day in winter or early spring, but very warm days during colder months are not unheard of and have been recorded since at least the early twentieth century. For example, some atypical January temperatures include 84°F (29°C) in Las Animas, Colorado, in 1916; 79°F (26°C) in Choteau, Montana, in 1919; and 82°F (28°C) in Indianola, Nebraska, in 1894. More recently, in January 1997, Zapata, Texas, had a record 98°F (36.7°C); and on March 31, 1998, Baltimore, Maryland, reached 97°F (36.1°C), and Concord, New Hampshire, saw 89°F (32°C). Probably the hottest December day recorded in North America was in La Mesa, California, which saw the thermometer climb to 100°F (37.8°C).

What are some of the **highest temperatures recorded**?

Record-High World Temperatures

Place	Year	Temperature (°F/°C)
El Azizia, Libya	1922	136/57.5
Death Valley, CA	1913	134/56.7

What is a Bermuda High?

A Bermuda High is the term given to an anticyclonic system in the western Atlantic that brings warm, humid air to the eastern seaboard of the United States. Occurring during the summer, a Bermuda High can last several weeks before dissipating.

Place	Year	Temperature (°F/°C)
Tirat Tsvi, Israel	1942	129/53.9
Oodnadatta, Australia	1960	123/50.6
Seville, Spain	1881	122/50
Rivadavia, Argentina	1905	120/48.9
Tuguegarao, Philippines	1912	108/42.2

How **warm** can it get in **Antarctica**?

Antarctica is not always an icebox perpetually below freezing. During the summer months, it can often get into the 40s and 50s Fahrenheit (4 to 14°C). The warmest temperature on record thus far on the continent is 59°F (15°C) at Vanda Station on January 5, 1974.

What is a **heat burst**?

In a manner similar to what happens when a Santa Ana wind rushes down into the low coastal areas of southern California, a heat burst raises temperatures by compressing air. Usually associated with thunderstorms, heat bursts typically occur during the night when turbulent air from as high up as 20,000 feet is forced downward as a storm begins to break up and dissipate. Meteorologists theorize that this happens when virga (rain that evaporates before reaching the ground) is present in dry, cold air. The resulting denser air is pulled down by gravity, increasing air compression near the ground rapidly and causing temperatures to rise unexpectedly.

In Waco, Texas, on June 15, 1960, it was reported that a heat burst caused the temperature to soar to 140°F (60°C) for a brief period that was accompanied by winds of about 80 miles (129 kilometers) per hour. More recently, in May of 1996, temperatures rose from about 88°F (31°C) to 101°F (39°C) in the towns of Chickasha and Ninnekah in about 30 to 40 minutes; and on August 3, 2008, Sioux Falls, South Dakota, saw the temperature shoot up from about 72°F (22°C) to 101°F (39°C) during gusting winds of 50 to 60 miles (80 to 100 kilometers) per hour. Perhaps most remarkable was a heat burst during the recent heat wave and fires in Australia. On January 29, 2009, at 3:00 A.M., a heat burst maxed out thermometers at 107°F (41.7°C).

What is **heat exhaustion**?

When the temperature is too high and people do not drink enough fluids, exert themselves too much, or spend too much time in the direct sunlight, heat exhaus-

Can weather cause bales of hay to catch fire?

Dry bales of hay are perfect tinder to start a fire. The most likely scenario for a hay bale to catch fire is when it is struck by lightning during a thunderstorm. But green, freshly harvested hay can be incendiary, too. Methane gas can build up within the piles of hay as the foliage decomposes, and when the temperatures are hot enough the methane can combust. This happened, notably, in the summer of 1995, when hay bales ignited in several U.S. states, especially in Missouri.

tion can result. Symptoms can include weakness, cold and clammy skin, paleness, fainting, irregular heartbeat, and vomiting. In extreme cases, heat exhaustion has been known to kill. There have been a number of cases where people exercising at a sports practice or conducting drills in the military have collapsed and died. Seniors and people who are already ill are particularly susceptible to heat exhaustion.

What is **heat stroke**?

Also called sunstroke, heat stroke happens when a person's body temperature exceeds 106°F (41°C). Symptoms include a rapid and strong pulse, hot, dry skin, and eventually unconsciousness. Heat stroke is considered a medical emergency, and immediate treatment is indicated to lower the body's temperature. Fluids, however, should be withheld. Hundreds of people die every year in the United States because of heat stroke, either while outside or while in an unconditioned room. Pets and livestock are also susceptible to this condition.

How **hot** can the **human body get** and still **survive**?

University-level research has included experiments that show that the human body can withstand a temperature of about 250°F (121°C) for up to 15 minutes. A more extreme case in which a professor conducted the research on himself showed that a healthy person could withstand 364°F (184°C) for one minute.

What are some other unusual events that saw a **sudden increase in temperature** in the United States?

Extreme shifts in weather patterns, such as when a strong warm or cold front moves through a region, have been known to cause rapid changes in temperature of 40 to 80 degrees Fahrenheit, sometimes in periods lasting less than an hour. One of the most stunning examples occurred in Great Falls, Montana, on January 12, 1980, when Chinook winds carried warm air into the city. A hydrologist for the National Weather Service reported that over a period of just seven minutes the temperature in Great Falls rose from –32°F (–35.5°C) to 15°F (–9.4°C), a change of 47 degrees Fahrenheit, or 22 degrees Celsius, which was a record, at

least in the United States. Previously, Montana held the record when on December 1, 1896, the temperature changed by 36 degrees Fahrenheit in seven minutes in the town of Kipp; it rose a total of 80 degrees Fahrenheit (37.7°C) that same day over just a few hours.

While not occurring over such an impressively short period of time, there have been many other incidents in the United States where abrupt temperature changes have happened within a relatively short time frame. Below are some other remarkable temperature changes.

Extreme Temperature Shifts in the United States within a 24-Hour Period

Location	Date	Temp Change (°F/°C)	Time Span
Loma, MT	January 14-15, 1972	103°F/48.6°C	24 hours
Granville, ND	February 21, 1918	83°F/39°C	12 hours
Kipp, MT	December 1, 1896	80°F/37.7°C	15 hours
Rapid City, SD	January 22, 1943	49°F/23°C	2 minutes
Great Falls, MT	January 12, 1980	47°F/22°C	7 minutes
Assiniboine, MT	January 19, 1892	42°F/19.8°C	15 minutes

COLD

What is the **lowest temperature possible**?

According to physics, the lowest temperature possible is zero degrees Kelvin (–459.67°F or –273.15°C), which is also known as absolute zero. This is the temperature at which molecular activity stops and it is not possible for any more heat energy to be lost. Such an extreme, of course, does not exist naturally on Earth, except in some university laboratories. The lowest air temperatures can go on the planet is about –130°F (–90°C); an unverified reading of –132°F (–91°C) was reported at Vostok, Antarctica, in 1997.

Who was **Paul A. Siple** and how is he associated with the concept of the **wind chill factor**?

The Antarctic explorer Paul A. Siple (1908–1968) coined the term in his 1939 dissertation, "Adaption of the Explorer to the Climate of Antarctica." Siple was the youngest member of Rear Admiral Richard Byrd's (1888–1957) Antarctica expedition (1928–1930) and later made other trips to the Antarctic as part of Byrd's staff and for the U.S. Department of the Interior assigned to the U.S. Antarctic Expedition. Siple later conducted experiments using a container of water subjected to specific temperatures and wind speeds to see how fast it would freeze. He also served in many other endeavors related to the study of cold climates.

Wind Chill Chart

Wind speed (mph)	Temperature (°F)																	
0	40	35	30	25	20	15	10	5	0	-5	-10	-15	-20	-25	-30	-35	-40	-45
5	36	31	25	19	13	7	1	-5	-11	-16	-22	-28	-34	-40	-46	-52	-57	-63
10	34	27	21	15	9	3	-4	-10	-16	-22	-28	-35	-41	-47	-53	-59	-66	-72
15	32	25	19	13	6	0	-7	-13	-19	-26	-32	-39	-45	-51	-58	-64	-71	-77
20	30	24	17	11	4	-2	-9	-15	-22	-29	-35	-42	-48	-55	-61	-68	-74	-81
25	29	23	16	9	3	-4	-11	-17	-24	-31	-37	-44	-51	-58	-64	-71	-78	-84
30	28	22	15	8	1	-5	-12	-19	-26	-33	-39	-46	-53	-60	-67	-73	-80	-87
35	28	21	14	7	0	-7	-14	-21	-27	-34	-41	-48	-55	-62	-69	-76	-82	-89
40	27	20	13	6	-1	-8	-15	-22	-29	-36	-43	-50	-57	-64	-71	-78	-84	-91
45	26	19	12	5	-2	-9	-16	-23	-30	-37	-44	-51	-58	-65	-72	-79	-86	-93
50	26	19	12	4	-3	-10	-17	-24	-31	-38	-45	-52	-60	-67	-74	-81	-88	-95
55	25	18	11	4	-3	-11	-18	-25	-32	-39	-46	-54	-61	-68	-75	-82	-89	-97
60	25	17	10	3	-4	-11	-19	-26	-33	-40	-48	-55	-62	-69	-76	-84	-91	-98

Wind Chill Chart

Wind speed (kph)	Temperature (°C)													
0	0	-1	-2	-3	-4	-5	-10	-15	-20	-25	-30	-35	-40	-45
6	-2	-3	-4	-5	-7	-8	-14	-19	-25	-31	-37	-42	-48	-54
8	-3	-4	-5	-6	-7	-9	-14	-20	-26	-32	-38	-44	-50	-56
10	-3	-5	-6	-7	-8	-9	-15	-21	-27	-33	-39	-45	-51	-57
15	-4	-6	-7	-8	-9	-11	-17	-23	-29	-35	-41	-48	-54	-60
20	-5	-7	-8	-9	-10	-12	-18	-24	-30	-37	-43	-49	-56	-62
25	-6	-7	-8	-10	-11	-12	-19	-25	-32	-38	-44	-51	-57	-64
30	-6	-8	-9	-10	-12	-13	-20	-26	-33	-39	-46	-52	-59	-65
35	-7	-8	-10	-11	-12	-14	-20	-27	-33	-40	-47	-53	-60	-66
40	-7	-9	-10	-11	-13	-14	-21	-27	-34	-41	-48	-54	-61	-68
45	-8	-9	-10	-12	-13	-15	-21	-28	-35	-42	-48	-55	-62	-69
50	-8	-10	-11	-12	-14	-15	-22	-29	-35	-42	-49	-56	-63	-69
55	-8	-10	-11	-13	-14	-15	-22	-29	-36	-43	-50	-57	-63	-70
60	-9	-10	-12	-13	-14	-16	-23	-30	-36	-43	-50	-57	-64	-71
65	-9	-10	-12	-13	-15	-16	-23	-30	-37	-44	-51	-58	-65	-72
70	-9	-11	-12	-14	-15	-16	-23	-30	-37	-44	-51	-58	-65	-72
75	-10	-11	-12	-14	-15	-17	-24	-31	-38	-45	-52	-59	-66	-73
80	-10	-11	-13	-14	-15	-17	-24	-31	-38	-45	-52	-60	-67	-74
85	-10	-11	-13	-14	-16	-17	-24	-31	-39	-46	-53	-60	-67	-74
90	-10	-12	-13	-15	-16	-17	-25	-32	-39	-46	-53	-61	-68	-75
95	-10	-12	-13	-15	-16	-18	-25	-32	-39	-47	-54	-61	-68	-75
100	-11	-12	-14	-15	-16	-18	-25	-32	-40	-47	-54	-61	-69	-76
105	-11	-12	-14	-15	-17	-18	-25	-33	-40	-47	-55	-62	-69	-76
110	-11	-12	-14	-15	-17	-18	-26	-33	-40	-48	-55	-62	-70	-77

At McMurdo Station, Antarctica, a bust of Rear Admiral Richard Evelyn Byrd commemorates the famous explorer and aviator. (*photo by Michael Van Woert, NOAA NESDIS, ORA*)

How does the **wind chill factor** affect temperature?

Also known as the wind chill index, it is a number that expresses the cooling effect of moving air at different temperatures. The National Weather Service began reporting the equivalent wind chill temperature, along with the actual air temperature, in 1973. For years it was believed that the index overestimated the wind's cooling effect on the skin. A new wind chill index was thus instituted in 2001, and other adjustments have since been made, too.

How is the **wind chill calculated**?

Wind chill is calculated using the following formula: Wind Chill = $35.74 + 0.6215T - 35.75(V^{0.16}) + 0.4275T(V^{0.16})$. The wind chill is in degrees Fahrenheit, T is the air temperature, and V is the wind speed in miles per hour. Above is the official wind chill chart from the National Weather Service as of 2009.

Does **drinking alcohol** truly **warm up** a person's **body**?

While drinking a shot of bourbon or other alcoholic drink on a winter's day might provide a sensation of being warmed up, in actuality alcohol makes a person's body more vulnerable to the cold. When you drink booze, what happens is that the alcohol causes blood vessels to constrict. This forces more blood to the surface of the skin, which stimulates nerves in the body and makes the drinker feel warmer. The actual effect, however, is that increased blood to the surface of the body causes one to lose more heat, lowering body temperature and making him or her more susceptible to frostbite. Add to this the fact that alcohol impairs one's judgment, and the idea of drinking on the ski slopes proves to be a very bad one indeed!

What is **frostbite**?

Frostbite occurs when people are exposed to cold temperatures for extended periods without sufficient protection. Skin and, if exposure is prolonged, deeper tissues, including nerves and blood vessels, can be damaged beyond repair. Frostbite begins at the extremities (fingers, toes, ears, etc.) as the body tries to protect vital organs by constricting blood vessels at the extremities. This moves warm blood toward the center of the body. While this natural adaptation is important for keeping the heart pumping and other organs working, if circulation is cut off too long in other areas, tissues will die.

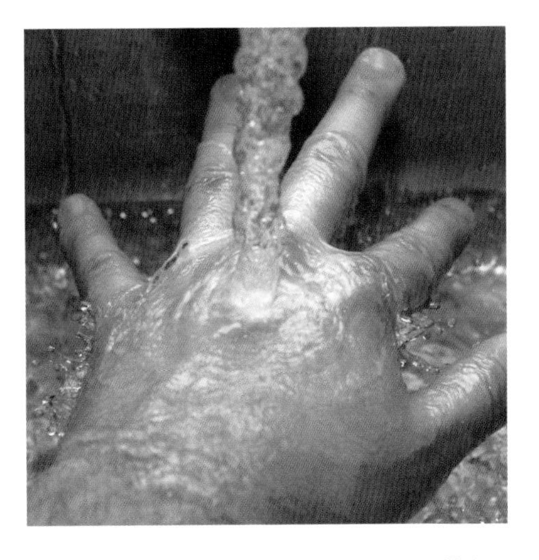

Always seek medical attention if you think you are suffering from frostbite. In a pinch, running warm water over the skin is one treatment if you can't get to a doctor right away.

The first stage of frostbite is frostnip, which is when skin begins to get numb but no damage to the tissues occurs. In first-degree frostbite, ice crystals begin to form on the skin, and a warming sensation in fingers or toes indicates the onset of second-degree frostbite. As the damage progresses to third-degree frostbite, extremities turn blue, white, or red, and in the last stages—fourth-degree—purple and then black. At this advanced stage, sensation is lost as nerves die, and it may be necessary to amputate. The best way to avoid frostbite is to simply stay indoors, especially if the wind chill is –50°F (–45.5°C) or lower. If you must go outside, bundle up, don't drink alcohol or caffeinated beverages, and also avoid smoking. Drinking and smoking can both constrict blood vessels, thus speeding the formation of frostbite.

What can I do to **help relieve frostbite**?

First of all, if you feel you are suffering from frostbite, you should seek professional medical attention while trying to keep pressure off of extremities (fingers and toes) that have been frozen. You should *never* vigorously rub the afflicted area as this could cause severe tissue damage or even break off pieces of tissue. If you are unable to see a doctor immediately, a warm bath (have the water between 100 and 110°F [38 to 43°C]) is a much better choice than sitting in front of a fireplace.

What are **chilblains**?

Chilblains refers to the burning, itching, redness, and chapping that can occur when skin suffers prolonged exposure to cold, damp weather conditions.

What is **hypothermia**?

Hypothermia is a life-threatening cooling of the body's temperature due to exposure to cold water, ice, or air. Chills and shivering can progress to much more serious symptoms, including disorientation, drowsiness, slurred or incoherent speech, memory loss, and unconsciousness. It only takes a decrease in core body temperature to about 94° to 95°F (34° to 35°C) for the onset of hypothermia to occur, and external conditions don't even have to be particularly harsh. For example, a human body floating in 40°F (4°C) water can suffer from hypothermia in less than half an hour. Professional medical assistance is, of course, indicated for treatment of

Is it possible to get frostbite in Hawaii?

Sure. Stand on top of Mount Haleakala on the island of Maui, or Mauna Kea on the Big Island, in the winter, or even spring or summer, and you can easily be exposed to temperatures well below –20°F (–29°C) for a long enough time to develop frostbite.

hypothermia, with the most important procedure involving stabilization of the body's temperature. An effective treatment is to have the patient inhale humid, warmed air (107° to 122°F [43° to 50°C]), which warms the respiratory and nervous systems and helps restore brainstem function that, in turn, controls heart function.

When adding **antifreeze** to a car, should you gauge the amount based on **expected low temperatures** or **expected low wind chills**?

After referring to your car's owner's manual, add the appropriate amount of antifreeze to protect your radiator at the lowest expected air temperatures for your region. You can research the Internet—sites such as www.weather.com are helpful—to find what the expected low is for your area. This generally works for urban areas, but if you live in a rural area you might wish to subtract an additional 20°F (9.5°C) to be safe.

Do **turtles and frogs** have **antifreeze** in their **blood**?

For a long time, it was a puzzle as to how cold-blooded reptiles and amphibians such as turtles and frogs survived in northern climes. These animals' body temperatures—unlike with warm-blooded animals such as dogs, deer, and people—plummet with the cooling temperatures. Many of them live underground near ponds and streams. As the temperatures outside cool, their body metabolism slows down, which makes it possible to survive long periods without much food, or even oxygen. Furthermore, many species of frogs and turtles have been found to use glucose compounds as a kind of antifreeze that allows their bodies to freeze without their blood and other body fluids forming harmful crystals that would otherwise destroy their tissues.

Why does my **nose run** when I'm **cold**?

When you inhale cold air, it causes the mucous membranes in your nose to constrict and then dilate. This reflexive action encourages the formation of mucous and a drizzly nose. After a while, however, the body adjusts to the colder temperatures and the sniffles should stop.

Can I **get a cold** from being out in **cold weather**?

The fact that it is cold will not, in itself, cause you to become sick. The reason why colds become more prevalent in winter is that people tend to stay indoors more, where they come into contact more frequently with other people, some of whom may be carrying viruses or illness-causing bacteria. Also, indoor heating systems run constantly—in colder regions, anyway—and the environment inside homes and buildings may not be all that healthy. Germs can proliferate in, for example, air filters and air ducts that have not been properly cleaned. Sometimes, too, people can be exposed to too much carbon monoxide, and the early symptoms of carbon monoxide poisoning can mimic those of a common cold.

To help avoid winter colds, it is wise to wash your hands regularly and, when in public areas, try and keep your hands away from your mouth, nose, and eyes. Recent studies have also shown that a good night's sleep (eight hours) helps to reduce the chances of getting a cold by 66 percent.

Some turtles (and frogs) have the ability to survive winters because their blood contains antifreeze-like glucose compounds.

How can a gardener **preserve unripe tomatoes** before a **winter frost**?

As any gardener knows, the first freeze of the winter season will kill many plants and ruin fruits and vegetables that have not yet been harvested. Tomatoes that are still a bit green can be placed in a paper bag and stored in a dry, dark pantry, where they should continue to ripen normally.

What methods do **farmers** use to **protect crops from frost and sudden freezes**?

Frosts and freezes can prove to be a major hazard to crops, especially in warmer climes where such incidents are rare. Some crops, such as citrus fruits, are particularly vulnerable, and there have been many years where orange, lemon, avocado, and other such produce have been nearly wiped out in places such as Florida, California, and Texas because of a sudden frost. One method that has been used to protect crops is called a smudge pot. These are portable burners that create heavy clouds of smoke, which help to hold in warmer temperatures closer to the ground. Environmentalists do not favor this method, however, because it creates a lot of pollution.

Another preventative strategy is to use huge wind machines. These giant fans blow warmer air downward onto the ground, circulating away the colder air that

Winters are an unpleasant time for most people, but some people are actually *allergic* to cold!

hangs low to the surface. Interestingly, farmers sometimes also coat their crops in ice to protect them from frost damage! Using sprinklers, they will spray the vegetation down, and, as long as the temperature does not drop below 25°F (–4°C), the crops do better than if they are allowed to go unprotected from the frost.

Why do more people die in **cold waves in the American South** than they do in the North?

Simply put, cold waves have proven more lethal in the South because people are not as well prepared for them. Homes and other buildings tend to have poorer insulation, and people do not own as many—if any—warm winter clothes.

What is the **coldest city** in America's **lower 48 states**?

Butte, Montana, holds the honor of being the coldest city in the contiguous 48 states. The temperature in Butte falls below the freezing point 223 days each year, on average.

What are some of the **coldest temperatures** measured in **Florida**?

The "Sunshine State" is not a complete stranger to wintry weather. For example, five inches (13 centimeters) of snow fell near what is now Jacksonville on January 11, 1800, and on February 13, 1899, the state had snowfall as far south as Ft. Meyers. Four inches (about 10 centimeters) of snow descended on Milton, Florida, on March 6, 1954. The farthest south that snow has been seen in Florida is in the city of Homestead, south of Miami, in January of 1977. There have been incidents of light snow flurries, sleet, and hail over the centuries, including recent snow flurries

Are some people allergic to the cold?

Research indicates that some people are, indeed, allergic to cold temperatures and may develop a rash as a result. Antihistamines can relieve symptoms, just as they do with exposure to other allergens.

in the Panhandle area on February 3, 2007, and flurries in Daytona Beach on January 3, 2008.

What is **permafrost**?

Permafrost refers to soil that is frozen all year long, even during the summer, for two consecutive years or more. Such conditions occur in areas where the mean annual temperature is 23°F (–5°C) or less. Construction in polar regions where permafrost is found can be problematic, because sometimes the ground can thaw beneath a building that was constructed over frozen ground; when this happens, building foundations are easily compromised.

Is there much **permafrost** in the **lower 48 U.S. states**?

There is no permafrost in the lower 48 states, though states such as North Dakota and Minnesota do often experience conditions where frost levels penetrate the ground to depths extended 30 or more inches (76 centimeters).

Have the **Great Lakes** ever been completely **frozen over**?

There have been occasions when the Great Lakes completely froze, which last happened in 1979. In the winter of 1993 to 1994, the lakes almost froze solid, but not completely. Even though this does not happen often, the Great Lakes generally shut down to shipping traffic during the months of January through March because the ice is far too hazardous.

Has the **Hudson River** ever **frozen over**?

The Hudson River in New York state extends about 300 miles (483 kilometers) from Lake Tear-of-the-Clouds on Mount Marcy down to New York Bay. When people think of the Hudson, what comes to mind is the part of the river that flows through the New York metropolitan area. This section does not freeze often, the last time being in January of 1918. However, at the higher elevations near the river's source, the Hudson freezes quite regularly.

What are some of the **lowest temperatures ever recorded**?

The table below lists some of the most chilling temperatures, by region, that have been recorded by meteorologists.

Is Moscow, Russia, colder than Minneapolis, Minnesota?

Moscow is at a latitude of 55° 45', while Minneapolis is at 44° 53'. Nevertheless, they have similar climates. Both cities experience mean January temperatures of about 14°F (–10°C), while Julys are about 66°F (18.9°C) in Moscow and 74°F (23.3°C) in Minneapolis on average.

Low Temperature Records

Place	Year	Temperature (°F/°C)
Vostok, Antarctica	1983	–129/–89.4
Klinck, Greenland	1991	–93/–69.6
Oimekon, Russia	1933	–90/–67.8
Verkhoyansk, Russia	1892	–90/–67.8
Northice, Greenland	1954	–87/–66
Snag, Yukon, Canada	1947	–81.4/–63
Ust'Shchugor, Russia	1978	–73/–58.1
Coyaique Alto, Chile	2002	–36/–37.7
Sarmiento, Argentina	1907	–27/–32.8
Ifrane, Morocco	1935	–11/–23.9
Charlotte Pass, New South Wales, Australia	1994	–9.4/–23
Mauna Kea Observatory, HI	1979	12/–11.1

What are some of the **coldest temperatures** ever recorded in the **United States**?

Coldest U.S. Temperatures

Location	Dates	Record Low (°F/°C)*
Prospect Creek, AK	January 23, 1971	–80/–62
Tanacross, AK	February 3, 1947	–75/–59.4
Rogers Pass, MT	January 20, 1954	–70/–56.7
Peter's Sink, UT	February 1, 1985	–69/–56.1
Riverside Ranger Station, MT	February 9, 1933	–66/–54.4
Moran, WY	February 9, 1933	–63/–52.7
Maybell, CO	February 1, 1985	–61/–51.7
Island Park Dam, ID	January 18, 1943	–60/–51.1
Tower, MN	February 2, 1996	–60/–51.1
Parshall, ND	February 15, 1936	–60/–51.1
McIntosh, SD	February 17, 1936	–58/–50
Tetonia, ID	February 9, 1933	–57/–49.4
Couderay, WI	February 4, 1996	–55/–48.3

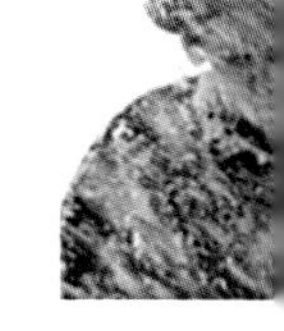

What is the best weather for maple syrup to be produced?

Maple syrup harvesters wait until nighttime temperatures dip below freezing, while day temperatures are above freezing. The conditions are ripe for the sap in maple trees to flow best.

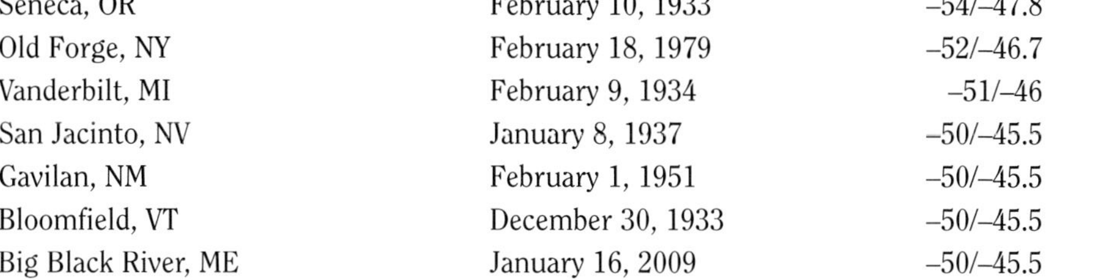

Location	Dates	Record Low (°F/°C)*
Seneca, OR	February 10, 1933	–54/–47.8
Old Forge, NY	February 18, 1979	–52/–46.7
Vanderbilt, MI	February 9, 1934	–51/–46
San Jacinto, NV	January 8, 1937	–50/–45.5
Gavilan, NM	February 1, 1951	–50/–45.5
Bloomfield, VT	December 30, 1933	–50/–45.5
Big Black River, ME	January 16, 2009	–50/–45.5

*Does not account for wind chill.

CLOUDS AND PRECIPITATION

CLOUDS

What is a **cloud**?

Clouds are collections of trillions of rain droplets suspended in the atmosphere. These droplets consist of water or ice crystals that condense around a nucleus—usually a dust or other small particle, such as pollen, volcanic ash, mineral flakes, or other organic or inorganic materials.

To form, clouds require air that has cooled to the point where the air is saturated with moisture or ice crystals. In other words, when the humidity reaches 100 percent. Each cloud droplet is only a few micrometers in diameter, and it take a million such droplets to form a single rain drop.

Why do **clouds float**?

Cloud droplets are, actually, subject to the Earth's gravity, and so they do, in fact, fall. They do so very slowly, however, and are subject to winds and air pressure that keep them aloft. When not being blown about, cloud droplets descend at a rate of about 30 feet per hour.

What are **supercooled droplets**?

A supercooled droplet is a cloud droplet that remains in its liquid form at temperatures far below the freezing point (as low as –40°F). Clouds formed of supercooled droplets pose potential hazards to aircraft that fly through them, because they can cause ice to form on wings and propellers. When low to the ground in the form of fog, supercooled droplets can also create hazardous road conditions by causing ice to form on streets and highways. On the other hand, supercooled droplets can also

Stratocumulus clouds are responsive to cloud seeding techniques. (*NOAA*)

result in great beauty when they paint the landscape in a sparkling layer of ice that makes trees and other plants shine.

What are **Aitken nuclei**?

The discovery that the air is full of tiny particles around which water droplets may condense to create clouds was made by Scottish physicist John Aitken (1839–1919). We usually just call them nuclei today, but, more formally, these particles are called Aitken nuclei.

How are **supercooled droplets** used for the purpose of **cloud seeding**?

Cloud seeding is a method for encouraging rain droplets to form in clouds. An experiment was conducted in 1946 by chemist and meteorologist Vincent Schaefer (1906–1993) and physicist, chemist, and Nobel Laureate Irving Languir (1881–1957) in which dry ice (frozen carbon dioxide) at –109°F (–78°C) was placed inside a cloud chamber containing cloud droplets chilled to –40°F (–40°C). The result was that rain droplets formed. The next step was to conduct the experiment in a real-life situation. In the winter of 1946, Schaeffer and Languir had dry ice placed in an airplane, which then dumped the load into a bank of stratocumulus clouds. They were pleased to see success when the dry ice caused a snow flurry to develop.

American physicist Bernard Vonnegut (1914–1997), who also worked with Schaefer, later experimented with other chemicals to see what else would be an effective way to seed clouds. He found that silver iodide, which is still used today, could perform the task well. Since then, other substances have been used to seed clouds, including sea salt and even water droplets. However, all these methods appear to work best in the winter. During warmer months, seeding clouds has proved to be more problematic because the correct conditions to produce precipitation are less common.

What is the **Bergeron-Findeisen mechanism**?

Swedish meteorologist Tor Harold Percival Bergeron (1891–1977) formed the theory that is accepted today about how clouds formed, including the role that ice crystals, water vapor, and temperature play. His theories were later confirmed by German meteorologist Walter Findeisen (1909–1945). The process of cloud formation is therefore known as the Bergeron-Findeisen mechanism.

A bank of cumulus clouds hovers over the Atlantic Ocean near Deerfield Beach, Florida. (*photo by Ralph F. Kresge, courtesy NOAA*)

How is **dry ice seeding** used to **increase visibility at airports**?

When the weather has been cold enough, using dry ice has proved to be effective in clearing fog around airports. Unfortunately, conditions are right for this method only about five percent of the time in the United States.

What is a **cloud chamber**?

Cloud chambers were originally designed to study radioactivity. Scottish physicist Charles Thomson Rees Wilson (1869–1959) invented the chamber in 1912, winning the Nobel Prize in physics in 1927 for his invention. The procedure involved saturating an enclosed chamber in water vapor until it was supersaturated. Ionized particles would then be passed through the chamber, serving as nuclei around which droplets would form. This had the advantage of making the particles visible to physicists, and the behavior of the particles could be studied.

Who was the first person to **classify clouds**?

The French naturalist Jean Lamarck (1744–1829) was the first to propose a system for classifying clouds in 1802. His work, however, did not receive wide recognition. A year later, the Englishman Luke Howard (1772–1864) developed a cloud classification system that has been generally accepted and is still in use today.

In Howard's system, he distinguished clouds according to their general appearance ("heap clouds" versus "layer clouds") and their height above ground. Latin names and prefixes are used to describe these characteristics. The shape names are *cirrus* (curly or fibrous), *stratus* (layered), and *cumulus* (lumpy or piled). The prefixes denoting height are *cirro* (high clouds with bases above 20,000 feet [6,000 meters]) and *alto* (mid-level clouds from 6,000 to 20,000 feet [1,800 to 6,000 meters]). There is no prefix for low clouds. Nimbo and nimbus is also added as a name or prefix to indicate that the cloud produces precipitation.

What are the **four major cloud groups** and their types?

Clouds are categorized in the following manner:

1. *High clouds* are composed almost entirely of ice crystals. The bases of these clouds start at 16,500 feet (5,000 meters) and reach up to 45,000 feet (13,650 meters).
   - A. *Cirrus* clouds (from the Latin for "lock of hair") are thin, feathery, crystal clouds that appear in patches or narrow bands.
   - B. *Cirrostratus* clouds are thin, white clouds that resemble veils or sheets. These clouds can be striated or fibrous in appearance. Because of the ice content, they are associated with the halos that surround the Sun or Moon.
   - C. *Cirrocumulus* clouds are thin clouds that appear as small, white flakes or cottony patches; they may contain super-cooled water.
2. *Middle clouds* are composed primarily of water. The height of the cloud bases range from 6,500 to 23,000 feet (2,000 to 7,000 meters).
   - A. *Altostratus* clouds appear as bluish or grayish veils or layers of clouds that can gradually merge into altocumulus clouds. The Sun may be dimly visible through them, but flat, thick sheets of these clouds can obscure the Sun.
   - B. *Altocumulus* clouds are white or gray and occur in layers or patches of solid clouds with rounded shapes.
3. *Low clouds* are composed almost entirely of water and may at times be super-cooled; at subfreezing temperatures, snow and ice crystals may be present as well. The bases of these clouds start near Earth's surface and climb to 6,500 feet (2,000 meters) in the middle latitudes.
   - A. *Stratus* clouds are gray, uniform, sheet-like clouds with a relatively low base, but they can also be patchy, shapeless, low gray clouds. Sometimes thin enough for the Sun to shine through, these clouds bring drizzle and snow.
   - B. *Stratocumulus* clouds are globular, rounded masses that form at the top of the layer.
   - C. *Nimbostratus* clouds are gray or dark, relatively shapeless, massive clouds that contain rain, snow, and ice pellets.
4. *Clouds with vertical development* contain super-cooled water above the freezing level and grow to great heights. The cloud bases range from 1,000 feet (300 meters) to 10,000 feet (3,300 meters).

The tops of cumulonimbus clouds look like giant cotton balls in the sky.

A. *Cumulus* clouds are detached, fair-weather clouds with relatively flat bases and dome-shaped tops. These usually do not have extensive vertical development and do not produce precipitation.

B. *Cumulonimbus* clouds are unstable, large, vertical clouds with dense boiling tops that bring showers, hail, thunder, and lightning.

Are there **other terms for describing clouds**?

Yes, in addition to the major naming conventions for clouds based on altitude and characteristics, there are also a variety of other Latin terms used to describe clouds. These names can be appended to the main names of clouds. For example, a cumulus castellanus is a cumulus cloud with formations on the top that look like castle towers. Below is a full list of other descriptors for clouds.

Cloud Type	Description
Arcus	Arched- or bow-shaped
Castellanus	Turret or tower-like
Congestus	Cauliflower-shaped
Duplicatus	Partly merged, double layered
Fibratus	Fibrous, filament-shaped
Floccus	Wooly, tuft-like
Fractus	Irregularly shaped
Humilis	Flattened, low, and small

Mammatus clouds over Tulsa, Oklahoma. (*NOAA Photo Library, NOAA Central Library; OAR/ERL/National Severe Storms Laboratory*)

Cloud Type	Description
Incus	Anvil-shaped
Intortus	Twisted and tangled
Lacunosus	Thin and with holes
Lenticular	Lens-like
Mammatus	Rounded, breast-shaped
Mediocris	Bulging, medium-sized (refers to cumulus clouds)
Nebulosus	Indistinctly shaped
Opacus	Thick
Pannus	Shredded
Perlucidus	Translucent
Pileus	Hood-shaped
Praecipitatio	Rain clouds
Radiatus	Lines of clouds radiating from a central point
Spissatus	Thick, grey cirrus
Stratiformis	Horizontal sheets
Translucidus	Transparent
Tuba	Curved, descending shape
Uncinus	Hooked shapes on the top of cirrus clouds
Undulatus	Wavy
Uniformis	Uniform
Velum	Sail-shaped
Vertebratus	Bone- or ribbed-like

What are **nacreous clouds**?

Nacreous clouds are clouds that occur at elevations of 12 to 20 miles (19 to 32 kilometers) high (rarely at lower altitudes) and look like cirrus or altocumulus lenticularis clouds. Often quite beautiful, they are sometimes called "mother-of-pearl clouds" because of irisation: supercooled water droplets causing refraction of sunlight that gives the edges of these clouds multiple colors in a kind of mother of pearl effect. Seen in northern climes, such as Alaska, Scotland, and Scandinavia, these clouds form only a couple hours before sunrise or after sunset.

What are **noctilucent clouds**?

Forming at altitudes of 47 to 56 miles (75 to 90 kilometers), these are the highest clouds you'll see in our atmosphere. Blown about by upper-atmosphere winds averaging 100 miles (161 kilometers) an hour, these cirrus-like clouds only form during the summer, and only at latitudes of 50 to 75 degrees north and 40 to 60 degrees south. They are usually seen at twilight and have a bluish or silvery color, sometimes flecked with red. It is speculated that noctilucent clouds may form as a result of meteor dust in the upper atmosphere because these clouds are more common when meteor activity increases.

What is a **mare's tail**?

More technically known as cirrus fibratus clouds, mare's tails get their name from their appearance. They are long, curved, and fibrous-looking.

What are the **cloudiest U.S. cities**?

In terms of annual average days when they were under overcast skies, the 10 cloudiest cities in the United States are as follows:

1. Astoria, OR, and Quillayote, WA: both 240 days
2. Olympia, WA: 229 days
3. Seattle, WA: 227 days
4. Portland, OR: 223 days
5. Kailspell, MT: 213 days
6. Binghamton, NY: 212 days
7. Beckley and Elkins, WV: both 211 days
8. Eugene, OR: 209 days

What kind of **cloud** has been **mistaken** for a **UFO**?

Altocumulus lenticularis (commonly called lenticular) clouds are sometimes called "flying saucer clouds" because they bear a strong resemblance to UFOs that have been reported over the years. These clouds, often appearing as one or more lens shapes stacked one on top of another, have been given many other names, too, including cap cloud, banner cloud, rotor cloud, crest cloud, foehn cloud, table

Sometimes, lenticular clouds are mistaken for UFOs because of their unusual disk shapes.

cloth, Chinook arch, Bishop wave, and Moazagotl. A peculiar property of these clouds is that they tend to remain stationary, even during winds gusting as much as 150 miles (241 kilometers) per hour. Typically, these clouds are formed near mountainous regions, where air is moving in a "standing wave" pattern. Moist air circulates above the cloud, and water vapor condenses, evaporating as it moves downwind and toward the ground.

What is a **mackerel sky**?

Mackerel skies are the result of altocumulus or cirrocumulus clouds forming a distinctive pattern that looks like the scales on a mackerel fish's back.

What is the **Table Cloth** cloud?

The Table Mountain near Cape Town, South Africa, is sometimes covered by a thin sheet of clouds when air is flowing off the top of the mountain in all directions. When this happens, the resulting cloud formation looks like a linen sheet; hence, local people have named it the Table Cloth.

How much of the **Earth** is usually **covered by clouds**?

At any given time, about one-half of the planet is covered by clouds.

How do **airplanes create clouds**?

When the air conditions are right and it's sufficiently moist, the exhaust from airplanes often creates condensation trails, known as contrails. Contrails are narrow lines of clouds that usually evaporate rather quickly. Contrails can turn into cirrus clouds if the air is close to being saturated with water vapor.

What is a **contrail**?

The word "contrail" is short for "condensation trail," and it refers to the water vapor that condenses around the exhaust of a jet aircraft flying at a high altitude. First studied during World War II, when there was a concern that contrails would give away the positions of B29 aircraft, contrails are now of interest to climatologists who worry about their effects on global warming. For instance, studies have shown that where contrails are present there is an increase in the formation of more cirrus clouds. Scientists are also concerned about the other chemicals in jet exhaust that could adversely affect chemical processes within the troposphere and lower atmosphere.

PRECIPITATION

What is **evapotranspiration**?

Evapotranspiration is the combination of water vapor being evaporated from the surface of the Earth (such as from lakes, rivers, or puddles) into the atmosphere, and transpiration, which is the movement of water from plants to the air.

How does the **hydrologic cycle** work?

The movement of water from the atmosphere to the land, rivers, oceans, and plants and then back into the atmosphere is known as the hydrologic cycle. We can pick an arbitrary point in the cycle to begin our examination. Water in the atmosphere forms clouds or fog and falls (precipitates) to the ground. Water then flows into the ground to nourish plants, or into streams that lead to rivers and then to oceans, or it can flow into the groundwater (underground sources of water). Over time, water sitting in puddles, rivers, lakes, and oceans is evaporated into the atmosphere. Water in plants is transpired into the atmosphere, too. The process of water moving into the atmosphere is collectively known as evapotranspiration.

What is **latent heat**?

The concept of latent heat was discovered by Scottish chemist Joseph Black (1728–1799), as well as, independently, by Swiss meteorologist, geologist, and physicist Jean André Deluc (1727–1817). When water condenses, cools, or freezes it loses energy and, thus, gives off heat. Water condensing into vapor releases an amount of energy equal to 600 calories per gram of water; when water freezes, it releases about 80 calories per gram. The heat released provides energy for storms to form.

What is a **"white-out"**?

An official definition for "white-out" does not exist. It is a colloquial term that can describe any condition during snowfall that severely restricts visibility. That may

How fast does rain fall?

The speed of rainfall varies with drop size, wind speed, and the size of the raindrops. A typical raindrop in still air falls about 7 miles (11 kilometers) per hour. Large raindrops can reach speeds of 16 to 20 miles (26 to 32 kilometers) per hour, while the tiniest drops are slower than a mile (1.6 kilometers) per hour as they drift to Earth. Thus, the larger drops can usually splash onto the ground about three minutes after being generated in a rain cloud that is about 5,000 feet (1,500 meters) above the surface.

mean a blizzard, or snow squall, etc. If you get some sunlight in the mix, that makes the situation even worse—it's like driving in fog with your headlights on high-beam. The light gets backscattered right into your eyes and you can't see.

What methods have been used to try to **remove hazardous fog**?

Helicopters have been used in the past to try and blow away fog; and in France people have used jet engines at airports to heat up the air. Both of these methods are very impractical, however.

Are there **different categories** of **rainfall**?

Yes. Rainfall is categorized into three types:

- *Convective rain* happens when the Sun warms the air near the ground; as the air then rises, it cools in the higher altitudes and water droplets form, creating a rain shower.
- *Orographic rain* is caused when air masses are elevated due to a geological feature such as a mountain. At the same time, land forms create a kind of squeegee effect on moisture as it runs into mountains. The result is the same as with convective rain, because as the air cools in the higher elevation, rain may result on the windward side of the hill or mountain.
- *Cyclonic rain* is the result of interacting air masses, which collide and force warm air masses upwards. This type of rain formation often results in strong thunderstorms or hurricanes.

What is the **shape of a raindrop**?

Although a raindrop has been illustrated as being pear-shaped or tear-shaped, high-speed photographs reveal that a large raindrop has a spherical shape with a hole not quite through it (giving it a doughnut-like appearance). Water surface tension pulls the drop into this shape. As a drop larger than 0.08 inch (two millimeters) in diameter falls, it will become distorted. Air pressure flattens its bottom and its sides bulge. If it becomes larger than one-quarter inch (6.4 millimeters) across, it will keep spreading crosswise as it falls and will bulge more at its sides, while at the same time, its middle will thin into a bow-tie shape. Eventually in its path downward, it will divide into two smaller spherical drops.

How **big** can a **raindrop** get?

The laws of physics restrain raindrops from getting too large before they break up into smaller droplets. Thus, about 0.25 inches (0.635 centimeters) is the largest a drop can get and still have the surface tension of the water hold it together.

A NOAA rainwater collector positioned near Mauna Loa, Hawaii, checks for acid rain. (*photo Commander John Bortniak, NOAA Corps*)

How is **rainfall measured**?

Agencies like the National Weather Service use very accurate devices that measure rainfall to the nearest one-hundredth of an inch. The devices, known as rain gauges or tipping-bucket gauges, collect rainwater at a point unaffected by local buildings or trees that may interfere with the rain.

What is meant by a **trace of precipitation**?

When the amount of rainfall is too little to be measured by a standard rain gauge, the precipitation is labeled a "trace."

How can I **measure** how much rain falls **where I live**?

Any container with a flat bottom and flat sides can measure rainfall. The width of the top of the container must be the same as at the bottom of the container, but the diameter does not matter. It could be a device purchased for measuring precipitation or something as simple as a coffee can.

What is the **Brückner cycle**?

The Brückner cycle refers to the idea that periods of unusually wet years are then followed by dryer-than-normal years in a cycle that alternates about every 35 years, though it may fluctuate by as little as 20 and as much as 50 years. It is named after German geographer and meteorologist Eduard Brückner (1862–1927). The cycle is also related to colder and warmer years. Brückner, who was also very interested in climate change and glacier advancement and retreats, based his theory on his research into glaciers and tree rings. Because there is so much variation in these

Can some people with arthritis or aching joints predict the rain?

Many people claim that they can feel a rain storm approaching because they will feel an ache in their knee, a throb in a tooth, or some other pain in their bodies. In a research study conducted by the University of Pennsylvania, scientists learned that, indeed, people with arthritis could sense when humidity levels went up and pressure dropped, both of which are good indications of an approaching storm.

cycles, though, climatologists have become much more interested in both the shorter-term and longer-term changes in climate.

Where does it **rain the most** in the **United States**?

Mt. Wai'ale'ale, on the island Kauai in Hawaii, receives a whopping average of 460 inches (1,168 centimeters) of rain a year—that's over 38 feet (about 12 meters) of rain per year!

What place has the **most rainy days** every year?

When it comes to the number of rainy days per year, the winner is again Mount Waialeale on Kauai, Hawaii, with up to 350 rainy days annually.

Where is the **rainiest place on Earth**?

The wettest place in the world, in terms of total rainfall, is Mawsynram, India, which drowns in 468 inches (1,188 centimeters) of rain each year, mostly because of monsoon rains. Nearby is Cherrapunji, India, which receives 460 inches (1,170 centimeters) of rain. Second place, though, goes to Tutunendo, Colombia. Here, the average annual rainfall is 463.4 inches (1,177 centimeters). Unofficially, Lloro, Columbia, endures 523 inches (1,328 centimeters) of rain annually, but no verified measurements have been taken to confirm this claim.

What are **ombrophobia** and **homichiophobia**?

Ombrophobia is an irrational fear of the rain, while homichiophobia is a fear of fog.

How much does it rain in the **Amazonian rainforest**?

The largest rainforest in the world surrounds the Amazon River basin, most of which lies within the borders of Brazil. Here, the average rainfall is 80 inches (200 centimeters) annually. Interestingly, despite all the rain and thick forest growth, the soil in the Amazon region is quite sterile and not well suited to farming.

Are stories about fish, frogs, and insects raining down from the skies real?

While the old saw about "raining cats and dogs" is just an expression, there have been reliable reports of very strange rains during which people are pummeled by frogs, grasshoppers, fish, and other bizarre creatures. For instance, in 1873 it was reported in a *Scientific American* article that frogs fell from the sky during a storm in Kansas City, Missouri. Both frogs and toads rained down on Minneapolis, Minnesota, during a 1901 storm; more recently, in 1995, a frog storm was reported in Sheffield, England.

One possible explanation for amphibian rain is waterspouts carrying frogs and toads up into the air, where prevailing winds then dump them onto a distant location. The same sort of theory might apply to reports of raining fish. A couple living in Folsom, California, for instance, reported a fish rain in September 2006, and earlier that year a similar eye-witness account came from Manna, India. Scientists were not incredulous, explaining that waterspouts can kick up winds of 200 miles (320 kilometers) per hour that have been known to lift objects as big as sailboats.

Other creatures, such as birds and flying insects, have been victims of weird weather, as well. It's not too much of a mental leap to conceive birds being caught in a strong storm that could injure or disorient them, causing them to fall out of the sky. Insects such as grasshoppers and crickets could just as easily be victims of such storms. In 1988, for example, meteorologists speculated that a swarm of red grasshoppers in Africa was caught up in strong winds and blown all the way to the Caribbean, where they landed in a massive insect rain shower.

Does it rain in the **Arabian Desert**?

Located on the Arabian Peninsula and covering an area of some 900,000 square miles (2,300,000 square kilometers), the Arabian Desert is, indeed, a very dry place, but rain does fall there. Some parts of this desert receive an average annual rainfall of a mere 1.38 inches (35 millimeters). On occasion, flash flooding occurs because of rainstorms. The worst of these occurred in 1995, when a storm and high winds caused flash floods that killed five people near Jiddah.

Does **all precipitation** reach the **ground**?

No. Rain and other precipitation can often evaporate before reaching the ground, especially when the air is dry (low humidity). In the American Southwest "dry" storms can be quite common, creating lightning and thunder but little precipitation. The danger of evaporating rain is that it can create conditions conducive to downdrafts and microbursts by cooling the air and causing changes in air pressure. Dry storms thus signal a warning to aviators about potential flying hazards.

Massive dust storms, such as this one near Stratford,Texas, on April 18, 1935, were a frightening sight that added to the woes of the Great Depression. (*photo by George E. Marsh Album, courtesy NOAA*)

What is **virga**?

Virga is a fancy name for rain that dries up before hitting the ground.

What is the **driest place on Earth**?

Probably the driest place on the planet is the Atacama Desert, which is located in Chile near the Pacific coast. The average rainfall here—specifically, in the town of Arica—is about 0.02 inches (0.05 centimeters). Meteorologists believe this never-ending drought is the result of the Humboldt Current, which blocks rain from reaching the Atacama Desert. There are some parts of the Atacama Desert that have not seen a drop of rain in centuries.

What **U.S. drought** has been the **costliest** in terms of financial damages?

There have been many severe droughts in the United States over the course of history. The most famous one is the Dust Bowl of the 1930s. However, in terms of money, the costliest drought thus far has been the 1988–1989 drought that resulted in an estimated $40 billion hit to the U.S. economy. Over half the country's population was negatively affected.

What was the **Dust Bowl** and what impact did it have on the United States?

The Dust Bowl drought and dust storms persisted, to varying degrees, from 1933 through 1939, devastating America's heartland. The worst drought years were in 1934 and 1939, and the worst dust storms occurred in 1935. The Dust Bowl turned

What happened because of a dust storm in 1977?

One of the worst dust storms to hit the United States since the Dust Bowl occurred in February 1977, when fierce winds blew over plowed fields from Colorado to Texas, kicking up a huge dust cloud. The dust storm dumped three million tons (2.7 metric tons) of soil onto the state of Oklahoma; the cloud continued through Mississippi and Alabama, dramatically raising particulate pollution, and continued across the country and far into the Atlantic Ocean.

once verdant farmlands into wastelands, and huge dust storms swept across Oklahoma, Texas, Kansas, Colorado, New Mexico, and even eastern states. But dry, hot weather was not the only culprit. Farmers at the time used techniques that depleted the soil severely. Most did not rotate their crops or irrigate their lands the way we do so today. The result was that when severe drought hit, crops died and the soil underneath was easily eroded. Strong winds blew the dirt into huge drifts, also sweeping away what little fertile soil was left.

The impact of the Dust Bowl was not only loss of crops, but also a never-before-seen period of immigration as former farmers abandoned their lands and, in many cases, headed out west to states such as California. Author John Steinbeck captured the plight of these people in his famous 1939 novel, *The Grapes of Wrath.* Photographers Dorothea Lange and Arthur Rothstein, among others, also recorded these terrible times for posterity in black-and-white photos while working for the Farm Security Administration. Today, the impact of the Dust Bowl can still be seen in America's central states. There remain many ghost towns that were once prosperous centers of commerce for local farm communities where the soil is still recovering.

What is **drizzle**?

Drizzle is just small droplets of rain measuring, on average, about 0.02 inches (0.05 centimeters) in diameter.

What place holds the record for the **least amount of rain** on Earth?

From October 1903 to January 1918, Arica, Chile, received no measurable rainfall.

What city in the **United States** holds the record for a **dry spell**?

From October 3, 1912, to November 8, 1914—a period of 767 days—no rain fell on Bagdad, California.

What does a **40 percent chance of rain** really mean?

When the morning weather report speaks of a 40 percent chance of rain, it means that throughout the area (usually the metropolitan area) there is a 4 in 10 chance

> **Why is it more likely to rain in a city during the week than on the weekend?**
>
> Urban areas have an increased likelihood of precipitation during the work week because intense activity from factories and vehicles produce particles that allow moisture in the atmosphere to form raindrops. These same culprits also produce warm air that rises to create precipitation. A study of the city of Paris found that precipitation increased throughout the week and dropped sharply on Saturday and Sunday.

that at least 0.001 of an inch of rain (0.0025 centimeters) will fall on any given point in the area.

HUMIDITY

What is **humidity**?

Humidity refers to the amount—or saturation—of water vapor in the air. Depending on air temperature and pressure, the air can contain differing amounts of humidity before the vapor turns into actual precipitation.

What is the difference between **absolute humidity** and **relative humidity**?

Absolute humidity is the *actual* amount of water that is mixed in with the air. It is measured in milligrams per liter. Relative humidity is expressed as a percentage and refers to the actual humidity divided by the maximum water vapor content possible at a given air temperature and pressure. For example: if a liter of air at 98°F (37°C) at one atmosphere pressure contains 44 grams (1.5 ounces) of water vapor, and the actual water content is 11 grams, then the relative humidity would be 25 percent ($11/44 \times 100 = 25\%$).

Can the **relative humidity** ever be **higher than 100 percent**?

No. It was once theorized that when clouds are in a state called "supersaturation" it was possible to have a relative humidity of slightly over 100 percent, but this has since been proven false.

What is **indoor humidity**?

Because we now live in homes with environments controlled by air conditioning, furnaces, and fans, the humidity inside is often very different from what the weather is like outside. For instance, in the winter many homes are very dry inside, which increases static electricity and causes uncomfortably dry skin. Overly humid air within a home can lead to the development of mold and mildew, some of which can be a risk to your health. Also, air that is too dry or too moist can cause structural damage to a house; overly moist wood, too, encourages pests such as termites to chew wooden supports and flooring. A comfortable humidity level for most people is between 30 and 60 percent

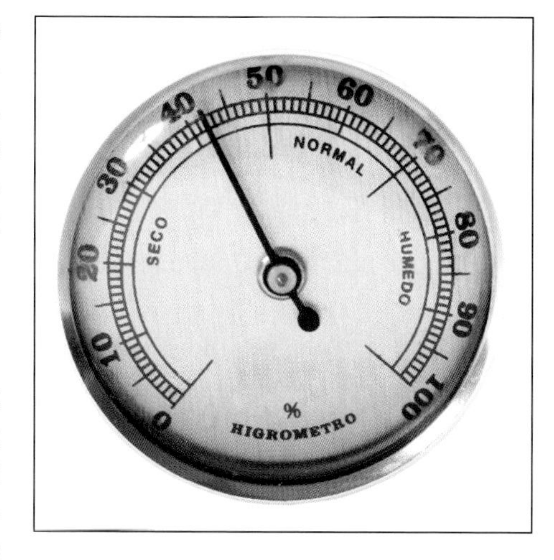

Hygrometers like this one have been used for hundreds of years to measure humidity.

What is a **hygrometer**?

A hygrometer is an instrument that measures the humidity. Italian artist and inventor Leonardo da Vinci (1452–1519) is often credited with constructing the first hygrometer; his design was later improved upon by Francesco Folli (1624–1685) in 1664, and Swiss physicist and geologist Horace Bénédict de Saussure (1740–1799) designed the first mechanical hygrometers. There are basically two types of hygrometers: 1) dry and wet bulb psychrometers, and 2) mechanical hygrometers. The first type uses dry and wet bulb thermometers to compare temperature changes resulting from humidity; the second type uses either organic material (blonde human hairs), which expand or contract based on humidity levels, or semiconductors made of lithium chloride or other substances whose degrees of electrical resistance change according to humidity.

Who **invented** the **dew point hygrometer**?

The dew point hygrometer (a type of dry and wet bulb psychrometer) was invented in 1820 by John Frederic Daniell (1790–1845). It consisted of two thin glass bulbs joined by a glass tube. One bulb held a thermometer and was filled with ether; the other was empty. As the air in the empty bulb cooled, the ether would condense on the thermometer, indicating the dew point temperature. Variations of Daniell's device are still used today, including cooled-mirror hygrometers that measure dew point when condensation forms on a mirror.

Is the **relative humidity** always **100 percent** during times when it is **raining or snowing**?

While the formation of rain and snow within a cloud requires humidity to reach a saturation point, by the time the precipitation reaches lower elevations and the ground the rain, hail, or snow can exist within air that has a lower humidity.

Who invented the **device** to measure **humidity**?

The hygrometer, which measures humidity in the atmosphere, was invented by the French physicist Guillaume Amontons (1663–1705).

What is humidifier fever?

If you do not clean your humidifier according to the manufacturer's recommendations, bacteria and mold may begin to grow on it, which could make you sick. Also, distilled water, not tap water, should be used in a humidifier because dissolved minerals in tap water can cause a fine white powder to blow in your home that can also encourage germs to spread. Though a variety of illnesses, ranging from allergic reactions to the common cold, could develop as a result, if they stem from a poorly maintained humidifier they might all be classified as "humidifier fever."

What is the **ideal relative humidity** that is the **most comfortable** for people?

A humidity level between 30 and 60 percent is generally considered comfortable for human beings, while keeping humidity below 50 percent has the added benefit of keeping dust mites under control in homes. Lower humidities tend to lead to dry or cracked skin, itching, and even respiratory problems, and higher humidity causes perspiration to be less effective in controlling body temperature, which makes people feel hotter. In northern climates, where winter dries out the air, humidity can drop below five percent, which is comparable to the humidity levels of a desert.

What is **dew**?

Dew is water vapor that has condensed onto cool surfaces. Usually, this happens overnight as air temperatures cool, and then the dew evaporates as the day progresses.

What is the **dew point**?

The dew point is the temperature at which air is full of moisture and cannot store any more. When the relative humidity is 100 percent, the dew point is either the same as or lower than the air temperature. If a fine film of air contacts a surface and is chilled to below the dew point, then actual dew is formed. This is why dew often forms at night or early morning: as the temperature of the air falls, the amount of water vapor the air can hold also decreases. Excess water vapor then condenses as very small drops on whatever it touches. Fog and clouds develop when sizable volumes of air are cooled to temperatures below the dew point.

What's the difference between **dew point, humidity,** and **relative humidity**?

While humidity and dew point are both measures of the amount of water contained in the air, humidity is measured as a percentage of water/air while dew point is indicated by temperature. The dew point indicates the temperature at which the air would become 100 percent saturated. The bigger the difference between the dew point and the current temperature, the less humid the air is. It is therefore the same type of measurement as relative humidity, except that relative humidity is expressed as a percentage and dew point is expressed using temperatures.

What is the difference between **absolute** and **specific** humidity?

Absolute humidity—given in grams per cubic meter—is the amount of water vapor in a cubic meter of air; specific humidity is the amount of water vapor per gram of air.

Does **dew fall**?

While this is a common expression, dew does not actually fall, a point that was proven in a demonstration conducted by Scottish-American physicist William Charles Wells (1757–1817) in 1814. Wells showed that dew actually formed on surfaces as water vapor condensed.

Where is the **most humid place** on Earth?

In terms of highest dew point temperatures, the extreme is found along the coast of Ethiopia along the Red Sea. Here, the average afternoon dew point in the month of June is a staggering 84°F (28.9°C). Tropical rainforests the world over, of course, are known for extremely high humidity.

ICE, SNOW, HAIL, AND FROST

How does **snow form**?

Snow forms in clouds in much the same way as water. Water vapor in the clouds collects around nuclei made out of dust or other particles in a cloud. When the temperature within the cloud is cold enough, and the water molecules begin to make contact with one another, they form crystals; and when the crystals become sufficiently heavy, gravity takes over and they fall to the ground.

The distinctive hexagonal crystal pattern of snowflakes occurs because of the six-fold molecular symmetry of ice. Water is composed of one oxygen bonding with two hydrogen atoms in a kind of V-shape configuration. When the temperature is cold enough and the molecules of water are drawn together, they naturally form into hexagonal rings. This pattern continues as the snowflake grows until it is apparent to the naked eye.

Does **water** always **freeze at 32°F (0°C)**?

The standard that water freezes at 32°F, or 0°C, is only true when the air pressure is exactly one atmosphere and when the water is fresh water and not salt water or some other form of water with impurities in it. Salt water (depending on the percent of salinity) freezes at a lower temperature than fresh water, and when the air pressure is greater than one atmosphere the melting point of fresh water is lowered. However, it takes a lot of pressure to make a difference: to lower the freezing point of water to 30°F (1°C) you must exert a pressure equal to 134 atmospheres!

Because water droplets in clouds must condense around impurities such as dust, they do not necessarily freeze at 32°F but can remain in a liquid state at temperatures as low as –40°F (–40°C). In other words, the water droplets cannot form ice until they begin to make contact with each other around a nucleus so they can form a crystal lattice. On the other hand, icebergs in the salty oceans are frozen at about the normal freezing point because icebergs are actually formed of fresh water. This happens because ice in the oceans is often formed slowly, allowing the ice crystals to force out impurities such as salt. On the other hand, ice flows—sections of the arctic ice sheet that break away each summer—are composed of sea ice; sea water begins to freeze at 28°F (–2°C) when there is no surface turbulence.

Another interesting property of ice on frozen lakes or in skating rinks is how friction makes ice skating possible. It was once believed that the pressure of an ice skate blade against the ice surface changed the state of the water from solid to liquid, thus creating a slippery surface on which to glide. More recently, it has been concluded that it is the friction of the blade against the ice that melts it and makes ice skating possible.

Why don't lakes and ponds **freeze solid** in the **winter**?

When ice forms in lakes, ponds, and other bodies of water it floats to the surface because liquid water is denser than solid ice. As water cools below 38°F (3.3°C) it becomes less and less dense, and thus more buoyant. This is good news for fish and other plants and animals that live in small bodies of water, since they can live below the layer of ice that forms.

What is **chionophobia**?

Chionophobia is a fear of snow.

What is the difference between **freezing rain, sleet,** and **hail**?

Freezing rain is rain that falls as a liquid but turns to ice on contact with a freezing object to form a smooth ice coating called glaze. Usually freezing rain only lasts a short time, because it either turns to rain or to snow. Sleet is frozen or partially frozen rain in the form of ice pellets. Sleet forms when rain falls from a warm layer of air, passes through a freezing air layer near the Earth's surface, and forms hard, clear, tiny ice pellets that can hit the ground so fast that they bounce off with a sharp click. Hail is a larger form of sleet.

What is the difference between a **frost** and a **hard freeze**?

The National Weather Service issues a hard freeze warning when it predicts that regional temperatures will fall below 27°F (–2.8°C) degrees for four hours or more.

Such a freeze is of particular concern to gardeners and farmers because temperatures this low will destroy crops. The good news is that a hard freeze usually spells the end for mosquitoes and other pesky insects for the rest of the year.

Frosts, on the other hand, do not require such low temperatures. In fact, air temperatures can be several degrees above freezing, while surface temperatures are at or below freezing, causing frost to form on car windows and other surfaces.

What are **glaze** and **rime**?

Glaze, as one might infer, is the result of freezing rain or drizzle falling onto surfaces and forming a sheet of ice. Rime, on the other hand, is formed by freezing fog or mist during windy conditions. There are two types of rime: hard rime and soft rime. Soft rime has a milky appearance and forms sugar-like crystals shaped into scales, feathers, or needles on the windward side of thin objects such wires, poles, and tree branches. Hard rime is less milky, comb-like in appearance; it is also more dense and less fragile than soft rime. Hard rime can form in windy weather when the temperature is between 18 to 28°F (–2 to –8°C) and humidity is 90 percent or more, while soft rime can occur in similar conditions, but when the temperature is below 18°F (–2°C).

What is an **ice storm**?

Ice storms account for some of the most dangerous winter conditions one can experience. Ice storms occur when freezing rain accumulates on the ground, building up layers of glaze or rime that coat everything, from roads to buildings to telephone lines and vegetation. Naturally, this makes driving extremely hazardous, and many people have lost their lives while traveling in ice storms. Airports will often cancel flights as ice forming on wings can prevent airplane wing flaps from moving, even when workers repeatedly de-ice them. Ice storms can down power lines and even cause older trees to collapse under the sheer weight of frozen water. For example, a tree 50 feet tall with a branch circumference averaging 20 feet can be weighed down by 10,000 pounds (about 4,500 kilograms) of ice as a result of an ice storm.

Rime ice coats tree branches near Asheville, North Carolina. *(photo by Grant W. Goodge, courtesy NOAA)*

How is **hail formed**?

Hail is precipitation consisting of balls of ice. Hailstones usually are made of concentric, or onion-like, layers of ice alternating with partially melted and refrozen snow, structured around a tiny central core. It is formed in cumu-

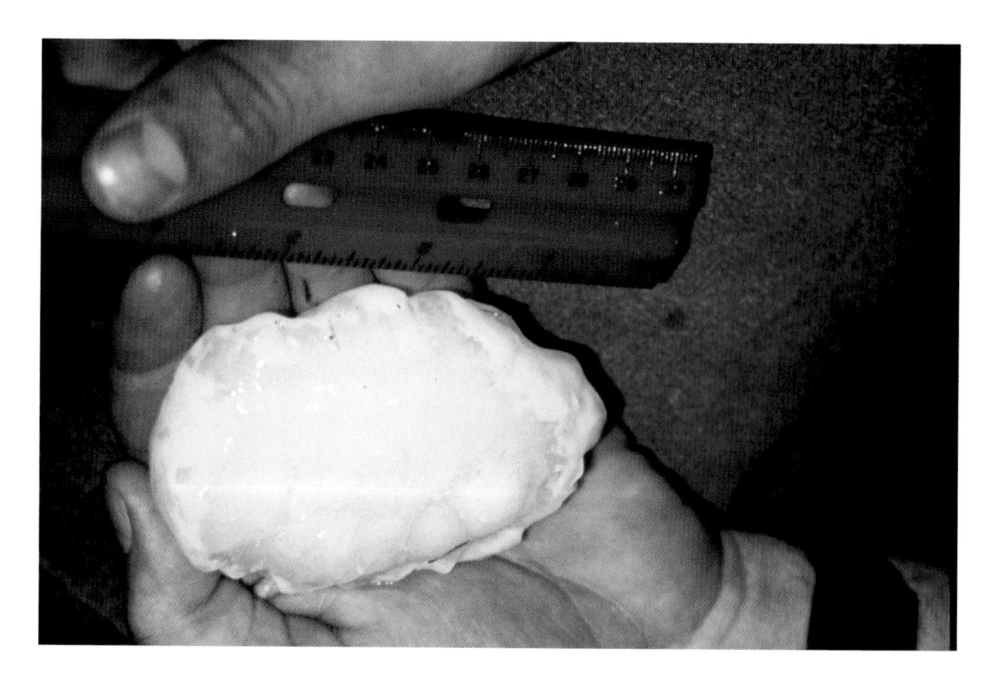

Texas is well known for its large hailstones, like this one that fell during a June 8, 1995, storm. *(NOAA Photo Library, NOAA Central Library; OAR/ERL/National Severe Storms Laboratory)*

lonimbus or thunderclouds when freezing water and ice cling to small particles in the air, such as dust. The winds in the clouds blow the particles through zones of different temperatures, causing them to accumulate additional layers of ice and melting snow and to increase in size.

Are **hailstones** always **round**?

Usually, hailstones are round or lumpy-round little ice balls. Sometimes, however, they can be oblong or have protruding spikes.

How large can **hailstones** become?

The average hailstone is about one-quarter inch (0.64 centimeter) in diameter. However, hailstones weighing up to 7.5 pounds (3.4 kilograms) are reported to have fallen in Hyderabad state in India in 1939, although scientists think these huge hailstones may be several stones that partly melted and stuck together. On April 14, 1986, hailstones weighing 2.5 pounds (one kilogram) were reported to have fallen in the Gopalgang district of Bangladesh; there was also a report of a hailstone weighing 4.5 pounds (2.04 kilograms) in Germany.

The current claim to fame for largest hailstone found in the United States was one that was measured to have a circumference of 18.75 inches (47.625 centimeters) that fell in Aurora, Nebraska, in June 2003. Before that, the record was held by a 17.5-inch (44.45-centimeter) stone found in Coffeyville, Kansas, in September 1970.

When it comes to weight, however, the Coffeyville stone was 1.67 pounds (0.76 kilograms) while the Aurora hailstone was a fluffier 1.3 pounds (0.59 kilograms).

What are the **different size categories** for **hailstones**?

In the United States, the following terms are used to describe the size of hailstones in weather reports. The actual sizes of the hailstones often don't match up to their supposed similes.

Hailstone Description	Diameter (inches/centimeters)
Pea	0.25 / 0.65
Marble	0.50 / 1.25
Penny, dime, large marble	0.75 / 1.90
Nickel, mothball	0.88 / 2.25
Quarter	1.00 / 2.50
Half dollar	1.25 / 3.20
Walnut	1.50 / 3.80
Golf ball	1.75 / 4.45
Hen egg	2.00 / 5.00
Tennis ball	2.50 / 6.35
Baseball	2.75 / 7.00
Tea cup	3.00 / 7.60
Grapefruit	4.00 / 10.25
Softball	4.50 / 11.40

Is there such a thing as a **megacryometeor**?

Yes. Megacryometeors are not hailstones because they don't form in clouds the way hailstones do. They are giant ice stones that do not require thunderstorms in order to form. Megacryometeors range in size from about a third of a pound (half a kilogram) to a monstrous one found in Brazil that weighed 137 pounds (62 kilograms). But how do they form? One argument is that they come from airplanes, but not from toilet drainage because this would mean contaminants and cleaners would be found in these large ice stones. However, ice might still form and break off of airplanes while in flight, and since it could take as much as three minutes for such a piece of ice to reach the ground, such megacryometeors might seem to come from out of the blue because the jet plane is long gone by then. A growing number of meteorologists, however, do not buy into the airplane scenario. They speculate that cooling and water vapor conditions within the tropopause that are still little understood may explain the formation of megacryometeors.

What are the **"hail belts"**?

Hail belts are regions that are ideal for hail storm formation. They can be found downwind of mountain ranges, usually in mid-latitude areas. Among the hail belt

Was the huge ice stone found in Hartford, Connecticut, really a hailstone?

No. On April 30, 1985, a 13-year-old-boy found a block of ice that was 1,500 pounds (680 kilograms), which had apparently fallen from the sky and onto his backyard lawn. No one ever figured out where it came from (sometimes, though, chunks of ice can fall from passing airplanes), but it was definitely not a natural hailstone.

regions are the Central Plains in the United States and Canada, Central Europe and parts of the Ukraine, southern China, Argentina, and parts of Central and South Africa and southeastern Australia.

What are some **notable record hail accumulations**?

In 1968, a storm in Illinois dumped the equivalent of 82 million cubic feet (2.32 million cubic meters) of ice over a 620,000 acre area. Then, in 1980, a hail storm in Orient, Iowa, left hail drifts that were six feet (1.8 meters) deep.

What **U.S. cities** tend to get the **most hail**?

The following U.S. cities are on the Top 10 list for receiving the most hail annually.

- Cheyenne, WY
- Tulsa, OK
- Amarillo, TX
- Oklahoma City, OK
- Wichita, KS
- Dallas/Ft. Worth, TX
- Arlington, TX
- Denver, CO
- Colorado Springs, CO
- Shreveport, LA

Do **hailstones** have **rings**?

Yes, but unlike with trees, the rings do not indicate the age of the hailstone. Italian physicist Alessandro Volta (1745–1827) did a study in 1806 to analyze the bands. He concluded that hailstones, before they fall to the ground, may change elevation a number of times due to up- and downdrafts. As the temperature and moisture levels change, layers of ice accumulate on an “embryo” (nucleus) at different rates, thus forming the layers.

Have **hailstones** ever proven to be **deadly**?

Strong hails can be heavy and plummet at speeds that have been known to break windows, dent cars, damage roofs, destroy crops (to the tune of about a billion dollars annually in the United States alone), and, yes, even kill. In July 1953, for instance, 30,000 ducks were found dead after a hail storm in Alberta, Canada. In Montana, a July 1978 hail storm killed about 200 sheep.

In the United States, not many people have been killed as a result of hail. The last reported death was on July 30, 1979. A hail storm in Fort Collins, Colorado, killed an infant and injured about 70 other people. Storms in other, poorer regions of the world are more common, however, because poorly built shelters collapse, killing those inside. For example, in Sichuan Province, China, a fierce hail storm injured 9,000 people and killed 100 on March 22, 1986. A month later, that same year, 92 people died in Gopolganj, Bangladesh, with some hailstones reported to weigh over two pounds (three kilograms).

What is **clear ice**?

Clear ice, as the name suggests, is clear, amorphous ice forming on surfaces as a result of water drops falling and freezing when the temperature is between 27 and 32°F (–3 to 0°C).

What is **graupel**?

Graupel—also known as snow pellets—is a funny word that means soft hail. It is created when supercooled water causes rime to form around a snowflake nucleus. Graupel is heavier and more granular than regular snow. Because of this, large amounts of graupel forming on a hill or mountainside can create dangerous conditions suitable for avalanches.

When does **frost** form?

A frost is a crystalline deposit of small thin ice crystals formed on objects that are at freezing or below freezing temperatures. This phenomenon occurs when atmospheric water vapor condenses directly into ice without first becoming a liquid; this process is called sublimation. Usually frost appears on clear, calm nights, especially during early autumn when the air above the Earth is quite moist. Permafrost is ground permanently frozen that never thaws out completely.

Frost coats a window with beautiful crystalline patterns.

Besides rime frost, what **other types** of **frost** are there?

Different weather conditions can cause frost to form in various ways, sometimes with spectacularly beautiful results. The types of frost include:

Advection (wind) frost is frost that forms on the edges of plants and other objects. Advection frost is formed on the upwind side of objects during very cold winds.

Fern (window) frost gets its name from the fernlike patterns it forms on windows, especially windows that are not well insulated. Flaws in the glass's surface provide the nucleus needed for water vapors to form crystals, which then radiate outwards in intricate patterns.

Frost flowers are the result of a rare interaction between plants and the weather. When water inside a plant stem cracks or splits due to the cold, the water can escape and then freeze into flower-like shapes. Because they are so fragile, frost flowers usually break apart or melt within hours of forming.

Hoar (radiation) frost is formed on clear nights when surface objects are colder than the surrounding air. It appears as white, loosely organized crystals. Hoar frost may appear similar to rime, but unlike rime it is formed without the presence of mist or fog.

What is the difference between **snow and hail**?

Snow is water vapor that freezes in clouds before falling to the Earth. Hail is water droplets (raindrops) that have turned to ice in clouds.

How much **water** is contained in **snow**?

Depending on conditions—and as anyone who has had to shovel snow can attest to—snow can range from light and fluffy to heavy, dense, and slushy. As a general rule of thumb, however, every 10 inches of snowfall that accumulates on the ground would equal about an inch of rain if it all melted.

Is it true that no **two snowflakes** are **exactly the same shape**?

Some snowflakes may have strikingly similar shapes, but these twins are probably not molecularly identical. In 1986, cloud physicist Nancy Knight believed she found a uniquely cloned pair of crystals on an oil coated slide that had been hanging from an airplane. This pair may have been the result of breaking off from a star crystal, or were attached side by side, thereby experiencing the same weather conditions simultaneously. Unfortunately the smaller aspects of each of the snow crystals could not be studied because the photograph was unable to capture possible molecular differences. So, even if the human eye may see twin flakes, on a minuscule level these flakes are different.

When were huge snowflakes observed?

On January 28, 1887, in Ft. Keough, Montana, there was a snowfall that included flakes measuring a spectacular 15 inches (38.1 centimeters) across! Of course, these flakes were not individual crystals, but rather clumps of crystals sticking together to form large flakes. Not long after these stunning flakes were seen, Shirenewton, England, experienced a storm in 1888 where 3.75-inch (9.5-centimeter) snowflakes were seen.

Can the form of **snowflake crystals be predicted**?

Snowflake crystals come in several forms, including needle-shaped, platelike, capped columns, and feathery dendrites. Temperature and humidity levels determine which type of shape is formed, so, yes, if these conditions are known, the type of snowflake formed could be predicted. In natural conditions, this would of course be impractical, but laboratory conditions could be established to form particular types of snow crystals, if desired.

How **unhealthy** is **snow shoveling**?

Heart attack rates increase sharply during the winter months in northern climates because people who are older or are not very healthy get too much exercise shoveling snow. Because more men than women tend to shovel snow, about three fourths of winter fatalities after snow storms are men. Fifty percent of these men, too, are over 60 years of age. You should always ask a doctor if you are healthy enough to shovel snow

Where was the **most snowfall** ever recorded?

Washington State's Mt. Baker recorded the most snowfall in a single season: 1,140 inches (2,896 centimeters).

What is **diamond dust**?

Diamond dust, also known as "ice prisms," are tiny ice crystals that can form in the air on extremely cold days if the air contains enough moisture. The effect can be quite beautiful, as sparkling, barely visible crystals appear in mid-air on sunny days, catching the sunlight and, indeed, appearing as if they are tiny diamond chips wafting in the breeze.

Who created the first **artificial snowflake**?

Japanese physicist Ukichiro Nakaya (1900–1962) created the first artificial snowflakes at Hokkaido University in 1936. Nakaya, who was inspired by the photographs of Wilson A. Bentley (1865–1931), also devised a rather poetic snowflake classification system, which he described in his 1954 book, *Snow Crystals: Natural and Artificial.*

Wilson A. Bentley was famous for taking highly detailed photographs of snowflakes, such as these images taken in 1902. (*NOAA*)

Is there a **classification system** for snowflakes?

Humankind has an affinity for classifying just about anything, and that includes snowflakes. In 1951, the International Commission on Snow and Ice (yes, there was such a commission!) created a system for putting a name on each type of snowflake—a daunting task when one considers that no two flakes are alike.

What are the **types of snowflakes** as identified in the **International Classification System**?

The types of snowflakes have been described, officially, as follows.

- *Stellar Plates,* as the name indicates, are starlike flakes that are flat, distinctly hexagonal, with six broad arms.
- *Sectored Plates* are similar to stellar plates, but also have prominent ridges pointing to each of the six facets in the plate.
- *Double Plates* occur when two stellar plates are connected by a cap. Usually, one plate is much larger than the other.
- *Split Plates and Stars* happen when parts of two separate plates merge to form one plate that, if not closely inspected, looks like a single six-armed plate. For example, a partial plate containing two arms could merge with one that has four arms left, leaving a six-armed plate that appears like a complete single plate.
- *Simple Prisms*—tiny, six-armed, flat snowflakes that are hard to distinguish with the naked eye, but are a very common form.
- *Stellar Dendrites* have treelike arms (*dendritic* means "treelike") with multiple branches extending from each of the six arms.
- *Fernlike Stellar Dendrites* are stellar dendrites with more frilly, fernlike branches.
- *Radiating Dendrites (Spacial Dendrites)* are dendrites that have arms extending not just in two dimensions, but in three.
- *Capped Columns* look like columns that have six flat sides (imagine two hexagons that are joined together).
- *Hollow Columns* are similar to capped columns, except the ends of the columns are hollow or divoted.
- *12-sided Snowflakes* When two six-sided plates join together at a 30-degree angle, they form what appears to be a 12-armed plate or capped column.

Who was the "Snowflake Man"?

American photographer and farmer Wilson A. Bentley (1865–1931) was nicknamed the "Snowflake Man," or just "Snowflake" Bentley, because he photographed images of over 2,400 snowflakes. His stunning photo collection capturing the natural beauty of snowflakes was published in 1931's *Snow Crystals.*

- *Needles* look just like the name: thin, long ice crystals. They usually form when the temperature is about 23°F (–5°C).
- *Triangular Crystals* often form at temperatures of about 28°F (–2°C) and are rather like deformed stellar flakes where half the arms are not fully formed, creating a triangle shape as a result.
- *Bullet Rosettes* happen when several columns melt and freeze together, looking like several crystal bullets merged at the heads at odd angles to each other.
- *Rimed Crystals* occur when additional water droplets freeze onto already formed snowflakes, giving them a fuzzy appearance.
- *Irregular Crystals* are snowflakes that are a rather disorganized mess of multiple snowflakes that have broken up and melted together.

What are some **record snowfalls** over a **24-hour period** in the United States?

There have been a number of impressive snowfall records in the United States since the 1990s, including the following:

Record One-Day Snowfalls

Place	Date	Snow (inches/cm)
Valdez, AK	January 15, 1990	47.5/120.6
Deadwood, SD	November 24, 2008	43.6/110.7
Buffalo, NY	December 9-10, 1995	37.9/96.3
New York, NY	February 11-12, 2006	26.9/68.3
Glasgow, MT	October 12, 2008	12.8/32.5

What is the **record** for the **greatest snowfall** in the United States?

The record for the most snow in a single storm is 189 inches (480 centimeters) at Mount Shasta Ski Bowl in California from February 13–19, 1959. For the most snow in a 24-hour day, the record goes to Silver Lake, Colorado, on April 14–15, 1921, with 76 inches (193 centimeters) of snow. The year record goes to Paradise Ranger Station, Mount Rainier, in Washington with 1,224.5 inches (3,110 centimeters) from February 19, 1971, to February 18, 1972; Paradise also has the highest average annual snowfall with 680 inches (1,727 centimeters). In March 1911, Tamarack, California, had the deepest snow accumulation—over 37.5 feet (11.4 meters).

Snow fences like this one are used to prevent snow drifts from encroaching on buildings.

Has it ever **snowed in Arizona**?

Higher elevations in the state actually receive quite a bit of snow. Flagstaff, for instance, is situated at an elevation of about 7,000 feet (2,133 meters), and experiences very cold temperatures in the winter (in January 22, 1937, the temperature dipped to –30°F [–34.4°C]). Though the climate of Flagstaff is officially "semi-arid," the city receives 100 inches (254 centimeters) of snow annually. Even in lower elevations, however, snow is not unknown in Arizona. Tucson received an extraordinary 6.4 inches (16.5 centimeters) of snow on November 16, 1957. This, of course, was not typical for the otherwise very hot city.

What are some other **unexpected places** where it has **snowed in the United States**?

In Louisiana, two feet (61 centimeters) of snow fell on the town of Rayne on February 14, 1895. Another old record is held by Savannah, Georgia, where 18 inches (46 centimeters) of the white stuff fell on January 10, 1800. More recently, residents of New Orleans were surprised to see snow accumulations on their front lawns on December 11, 2008.

What is a **heavy snow warning**?

A heavy snow warning is issued when the National Weather Service expects an accumulation of four inches (10 centimeters) or more within a 12-hour period. Heavy snow warnings differ from blizzard warnings in that they do not depend on strong winds for an advisory to be issued.

What is the **purpose of a snow fence**?

Snow fences are usually installed in places where there is not much in the way of vegetation or buildings to keep winds from blowing and drifting snow. The fences create wind turbulence that prevents drifting downwind of the fence.

What is a snow roller?

People aren't the only ones who enjoy building snowmen. Sometimes, nature gets into the game as well. In windy, wintry conditions, breezes have been known to start small collections of snowflakes rolling. As they roll, snow accumulates, and the snowball gets bigger and bigger. Such snow rollers have been known to grow to diameters of several feet.

What are the **10 snowiest cities** in the **United States**?

The annual average snow precipitation for the 10 U.S. cities that have the most snow is listed below.

Top 10 Snowiest U.S. Cities*

City	Average Annual Snowfall (inches/cm)
Truckee, CA	203.4/516.6
Steamboat Springs, CO	173.3/440.2
Oswego, NY	153.3/389.4
Sault Ste. Marie, MI	131.2/333.2
Syracuse, NY	120.2/305.3
Marquette, MI	118.2/300.2**
Meadville, PA	111.2/282.4
Flagstaff, AZ	111.1/282.2
Watertown, NY	110.8/281.4
Muskegon, MI	105.9/268.9

* Cities included have a population of 10,000 people or more; statistics are averages from 1971 to 2000.

** The Marquette airport gets 179.8 inches (456.7 centimeters) on average.

Is it ever **too cold** to snow?

No matter how cold the air gets, it still contains some moisture, and this can fall out of the air in the form of very small snow crystals. Very cold air is associated with no snow because it is usually very dry and these invasions of air from northerly latitudes are associated with clearing conditions behind cold fronts. Heavy snowfalls are associated with relatively mild air in advance of a warm front or on the back side of a strong low pressure system. The fact that snow piles up, year after year, in Arctic regions illustrates that it is never too cold to snow.

Can you use **snow** to keep you **warm**?

Snow, just like wood or stone or soil, can be used as an insulator. If there is sufficient snow on the ground, one can dig out a small cave or igloo-like structure, crawl inside of it, and wait for one's body heat to warm the inside. To work more efficiently, it is best to be six or more feet (two meters) below the surface, and keep the opening of your snow cave downwind. Such makeshift structures can warm up to 60°F (15°C) or more, and can be quite cozy. They are a good way to avoid exposure if one has become lost or stranded outdoors in the winter.

How **warm** can the weather be and yet have **snowfall**?

It is possible for the temperature on the ground to be in the 40s Fahrenheit (4 to 9°C) and still snow. This happens when snow forms at colder temperatures in clouds and does not melt before reaching the ground. In one case, a temperature of 47°F (8°C) was measured at New York City's LaGuardia Airport when snow was also seen.

What is the **snow line**?

The term snow line can refer to two different things: 1) the elevation on a hill or mountainside above which snow has fallen and below which the precipitation turns to rain; or 2) the latitude north of which is covered by snow.

What are you seeing when you "**see your breath**"?

It can be fun for kids to breathe out on a cold day and pretend that they are perhaps dragons, but it is obviously not smoke coming out of their mouths: it is water vapor. What is happening is that the moisture in one's breath (humidity) is turning to fog as it leaves the warm confines of the mouth and hits the chilly air. While one or two people breathing out on cold days will not affect the weather, it has been observed that large herds of animals huddling together on a winter's day can actually produce a small fog bank.

What places experience the **most annual precipitation** on Earth?

Highest Annual Precipitation

Place	Precipitation (inches/cm)	Years on Record
Lloro, Columbia	523.6/1,330.9*	29
Mawsynram, India	467.4/1,187.2	38
Mt. Waialeale, Kauai, HI	460/11,68.4	30
Debundscha, Cameroon	405/1,028.7	32
Quibdo, Columbia	354/899.2*	16
Bellenden Ker, Queensland, Australia	340/863.6	9
Henderson Lake, British Columbia, Canada	256/650.2	14
Crkvica, Bosnia-Herzegovina	183/464.8	22

*Estimated.

What is the **most rainfall ever recorded** within a one-day period?

Tropical Storm Claudette brought a record U.S. rainfall of 43 inches in 24 hours in and around Alvin, Texas, in July of 1979.

The coastline near Arica, Chile, known as the driest place on Earth.

What places have the **lowest annual precipitation** on Earth?

Lowest Annual Precipitation

Place	Precipitation (inches/cm)	Years on Record
Arica, Chile	0.03/0.076	59
Place	**Precipitation (inches/cm)**	**Years on Record**
Amundsen-Scott Station, Antarctica	0.08/0.2	10
Wadi Halfa, Sudan	< 0.1/0.025	39
Batagues, Mexico	1.2/3.05	14
Zaranj, Afghanistan	1.36/3.45	N/A
Aden, Yemen	1.8/4.57	50
Mulka, South Australia	4.05/10.29	42
Astrakhan, Russia	6.4/16.26	25
Puako, HI	8.93/22.68	13

What **country** can claim to have the **worst weather**?

That redoubtable distinction goes to the United States of America. Subjected to hurricanes, flooding, drought, heat and cold waves, blizzards, and the worst tornado activity on the planet, the United States experiences more weather disasters than any other nation.

How much have **disasters cost** in recent years to the United States in terms of money and deaths?

Since 1990, natural disasters have cost the United States about $540 billion and nearly 5,000 lives. This period of history unfortunately includes the 2005 Hurricane Katrina disaster, which is responsible for 1,833 deaths and almost $134 billion in damages just by itself. More typically, annual damages inflicted on American soil average about $18 billion annually. This includes damages caused by hurricanes, tornadoes, droughts, floods, blizzards and ice storms, heat waves, wildfires, and other severe conditions. The table below breaks this down by year.

Costs of U.S. Weather Disasters, 1990–2008

Year	Cost in Billions*	Deaths
1990	$7.1	13
1991	$6.2	43
1992	$45	87
1993	$40.9	338
1994	$8.4	81
1995	$18.6	99
1996	$18.7	233
1997	$10	114

Year	Cost in Billions*	Deaths
1998	$27.7	399
1999	$12.2	651
2000	$7.2	140
2001	$7.8	46
2002	$15.6	28
2003	$14	131
2004	$49.5	168
2005	$171.2	2,002**
2006	$11.8	95
2007	$10.9	22
2008	$56.7	274

* Costs are in 2008 U.S. dollars, adjusted for inflation.
** Includes Hurricane Katrina.

What is **emergency response planning**?

Most disasters in the United States (about 85 percent) occur because of the weather, with the rest due to such geologic events as volcanic eruptions and earthquakes, or acts of terrorism. In an effort to minimize the costs of such tragedies, a system connecting federal, state, and local agencies has been established to issue weather warnings, plan evacuation routes, and coordinate relief efforts. Community awareness programs, building code enforcement, and computer modeling are all employed in this effort, with meteorologists naturally playing a key role.

How have **weather-related forest fires** created tragic disasters in the **United States**?

Dry, hot conditions and lightning have caused some of the most devastating forest fires in the United States. For example, on September 1, 1894, a heat wave in the Midwest led to a forest fire near Hinkley, Minnesota, where 400 people consequently lost their lives; and on October 8, 1871, the so-called Peshtigo fire killed 1,800 people in Michigan and Wisconsin. Northern climes are particularly vulnerable to forest fires caused by the weather. During Alaska's mild, dry summers, for instance, the state experiences a dramatic rise in forest fires.

How do some **animals** appear to be able to **predict an oncoming storm**?

There is a considerable amount of folklore and old wives tales about how to predict the weather based on animal behavior. On the other hand, there do, in fact, seem to be ways one can tell if a storm is approaching, just by observing animals. Biologists and other scientists, as well as other people who work with animals daily, have noted some truth to the following:

- Geese tend not to fly when a storm is nearing. One explanation for this is that air pressure drops when foul weather approaches, and it might be harder for some larger birds, such as Canada geese, to get off the ground in such conditions. More likely, though, is that the birds instinctively understand that lower air pressure is a good indication of an upcoming storm.

- Seagulls and other ocean birds also tend to remain grounded before bad weather ensues.
- Many farmers believe that cows stay away from hilltops and remain close together in herds. Similar behavior has also been observed among deer and elk.
- Because frogs like it moist and humid, which tends to be the case during and just before stormy weather, they remain out of the water for longer periods of time. You can tell because they are croaking and ribbiting a lot more before it rains.
- Mosquitoes and biting black flies tend to bite and suck blood with much more ferocity and urgency before bad weather, perhaps saving up so they don't go hungry while they seek shelter from the storm.

BLIZZARDS AND AVALANCHES

What is a **blizzard**?

According to the U.S. National Weather Service, a winter storm is considered a blizzard when wind speeds reach 35 miles (56 kilometers) per hour and there is poor visibility of less than one-quarter of a mile (400 meters). Snow does not need to be falling at the time, but blowing and drifting should occur with drifts exceeding 10 inches (25 centimeters) deep.

How bad was the **Blizzard of 1888**?

After a severe blizzard hit the High Plains of the United States in February 1888, causing the deaths of many people and farm animals, an even more destructive blizzard wreaked havoc on the East Coast from Maine to Chesapeake Bay from March 11 to 14. Several feet of snow fell all over the region, and in Saratoga Springs, New York, 52 inches (1.32 meters) of snow fell and there were drifts of up to 52 feet (16 meters) deep. Wind speeds ranged up to 70 miles (113 kilometers) per hour. By the time the storm was over, more than 400 people had lost their lives.

How destructive was the **Blizzard of 1996**?

The Blizzard of 1996 left a path of destruction from Georgia to Pennsylvania. Almost $600 million in insurance claims were filed and 187 people lost their lives. Two or more feet of snow covered Washington, D.C., West Virginia, New England, and New York City. In addition, about $700 million in flood damage was reported in Pennsylvania. All together, this one storm cost $3 billion.

What should you do if you are **stranded in your car** during a **blizzard**?

If you are far from any buildings or people who might help, rather than getting out of your car and wandering off, possibly getting lost or freezing to death, it is always

Do loud noises trigger avalanches?

The notion that making loud noises, such as shouting or clapping one's hands, near dangerous snow conditions will trigger an avalanche is an old myth. Actually, it would take a tremendous sound, such as a nearby sonic boom where an avalanche is about to occur anyway, for noise to make a difference. In 90 percent of cases, avalanches are precipitated by a person's or persons' weight, or the weight of a snowmobile or other machine, on top of unstable snow.

a better idea to remain in your vehicle. Hopefully, you will have a charged cell phone handy, but if not it is still better to remain in the car. After checking to make sure that the car's exhaust pipe is not blocked by any snow or ice (being wary of carbon monoxide poisoning), keep the car running for as long as possible to stay warm. Also, turn on your hazard lights, which may attract help, such as a police officer, while also warning snow plows not to come too close. During the winter months, it is a good idea to keep blankets, food, a spare tire, and a first aid kit in your car. Never drink alcohol to try and stay warm, but a thermos containing a hot beverage or soup would prove handy.

What happens to **domesticated animals during blizzards**?

Farm and range animals certainly suffer as much as people during blizzards. For example, in the Blizzard of 1886 that hit the central plains states of America, about 80 percent of livestock perished in the worst-affected areas of Texas, Oklahoma, Nebraska, and Kansas. In 1966, 100,000 head of livestock died in Nebraska, Minnesota, North Dakota, and South Dakota when a blizzard dumped 30 feet (9 meters) of snow, drifts of up to 30 feet (over 9 meters) accumulated, and winds gusted up to 100 miles (160 kilometers) per hour.

What **causes avalanches**?

The most dangerous conditions for an avalanche occur when a lot of snow has fallen and/or blowing wind has caused snow to accumulate within a short period of time (hours or a few days, versus weeks). A "dry slab" avalanche is the most hazardous. This is when a heavy slab of snow that has formed quickly is resting above another layer of snow that is weaker but formed over a longer period of time. Dry slab avalanches are usually set off by a person walking over the unstable layer. There are also "wet slab" avalanches, which, as one might guess, involve a layer of wet snow over a harder layer of snow.

Avalanches are most likely to occur on hills with inclines of 30 to 45 degrees, though wet snow can tumble down a hill with a grade of as little as 10 degrees, and dry snow regularly causes avalanches on hills with about 20 to 22 degree slopes. Avalanches happen abruptly, and once a slab has broken off, there is usually no escape for someone downhill. Traveling at 60 to 80 miles (95 to 130 kilometers) per hour, an avalanche will quickly bury everything in its path.

Most avalanches are triggered simply by gravity acting on heavy snow banks, but sometimes a careless snowmobiler or skier may be the culprit.

What is the most **dangerous month** for **avalanches**?

More avalanches occur during the month of February in the United States than any other month. Colorado, Alaska, and Montana are the states where most avalanche-related deaths happen.

What is a **sluff**?

A sluff is a layer of loose snow. On rare occasions, an avalanche may be comprised of sluff rolling down a hill, but more frequently it is a dry or wet slab that causes an avalanche.

How many people **die in the United States** because of **avalanches**?

Deaths from avalanches are fairly uncommon, but they do occur, often as a result of carelessness or from not heeding posted signs. Below is a list of fatalities covering the last decade.

U.S. Avalanche Fatalities from 1998 to 2008

Season	Deaths
1998–1999	29
1999–2000	22
2000–2001	33
2001–2002	35
2002–2003	30
2003–2004	23
2004–2005	28
2005–2006	24
2006–2007	20
2007–2008	36

What is the **worst avalanche disaster** on record?

In 1970, an avalanche in Yungay, Peru, killed 20,000 people.

> **What unique weapon was employed during World War I?**
>
> During the War to End All Wars, avalanches were deliberately triggered in some battles in Europe as a weapon. Estimates are that between 40,000 and 80,000 soldiers died as a result.

What is the **Storm of the Century**?

Over the years, a number of storms have been called "storms of the century." The twentieth century experienced several storm events that could certainly qualify, or at least be nominated, for this honor. A huge blizzard struck the Midwest from January 10 to 11, 1975, that included snowfalls in Nebraska reaching 19 inches (48 centimeters) deep, wind chills of –80°F (–62°C) in the Dakotas, and wind bursts of 90 miles (145 kilometers) per hour in Iowa. Eighty lives were lost as a result of this storm.

Another candidate for the title arrived on stage in 1993, when a blizzard struck the American East Coast, killing 318 people, including 48 at sea. Fifty percent of the American population was affected in some way by the storm. The storm reached from Maine to Florida, where half a foot of snow even fell in the Florida Panhandle, and even Daytona Beach saw freezing temperatures. Winds near Key West raged at up to 109 miles (175 kilometers) per hour. Meanwhile, Mount LeConte, Tennessee, saw 56 inches (142 centimeters) of snow, and in Syracuse, New York, there was 43 inches (109 centimeters) of the white stuff.

The 1993 storm ranged far beyond U.S. borders, however, extending north to Canada and south all the way to Central America. At its peak, it reached the strength of a category 3 hurricane, and by the time it was over it had dumped 44 million acre-feet (about 14.3 trillion gallons, or 54.3 trillion liters) of water onto the ground. Add to this several killer tornadoes, and perhaps the 1993 storm wins the twentieth century's title as "storm of the century."

HURRICANES, MONSOONS, AND TROPICAL STORMS

What was the **Perfect Storm**?

The subject of a 1997 novel by Sebastian Junger, as well as a 2000 movie starring George Clooney, Diane Lane, and Mark Wahlberg, the Perfect Storm was also a real, terrifying event. During the last days of October 1991, an extratropical cyclone organized itself several hundred miles off the coast of Nova Scotia. At the same time, Hurricane Grace—a relatively weak category 2 hurricane—moved toward this storm from the south. As Grace approached the northern cyclone, the winds from the cyclone did something unusual. Typically, tropical hurricanes will tend to move away from the shoreline as they move north, but the cyclone winds, spinning in a northeastern swirl, blew Grace toward the Northeast Coast. The two storms merged and became known as the 1991 Halloween Nor'Easter.

The result was one of the most destructive storms in history, with winds up to 65 knots (75 miles per hour or 120 kilometers per hour) and ocean waves up to 39 feet (12 meters) high. A dozen people died in the storm (including the six crew members aboard the swordfishing vessel *Andrea Gail,* the subject of Junger's novel) and one billion dollars in damages were incurred. In another unusual twist, the Halloween Nor'Easter had turned into a hurricane by November 1, an event that is quite uncommon for an extratropical low pressure system, though not unheard of. The new hurricane, however, remained unnamed because the National Weather Service thought it might be confusing to the American media to rename the Halloween Nor'Easter.

What are **monsoons**?

Occurring in southern Asia, monsoons are winds that flow from the ocean to the continent during the summer and from the continent to the ocean in the winter. The winds come from the southwest from April to October, and from the northeast (the opposite direction) from October to April. The summer monsoons bring a great deal of moisture to the land. They cause deadly floods in low-lying river valleys, but also provide the water southern Asia relies upon for agriculture.

What is the **origin** of the word **"monsoon"**?

The word "monsoon" comes from the Arabic word "mausin," meaning season.

What is an **Alberta clipper**?

An Alberta clipper is a storm that develops on the Pacific front, usually over the Rocky Mountains of Alberta, Canada. This quick-moving storm moves southeast into the Great Plains, leaving a trail of cold air and snow.

What is a **Siberian express**?

This term describes outbreaks of arctic air that are severely cold; they descend from northern Canada and Alaska to other parts of the United States.

What is a **hurricane**?

A hurricane is a tropical storm that has winds of 74 miles per hour or more that forms in the Atlantic Basin. Hurricanes typically occur in the North Atlantic and Caribbean Sea during the months of July, August, and September, when warm surface ocean temperatures exceed 80°F (26.5°C), providing energy that feeds into the storm. Seawater evaporates into the air, creating clouds, while the Coriolis effect causes the clouds to rotate.

For a hurricane to develop, there must not be a lot of difference in wind speeds in the upper and lower elevations of the storm. If there is a big difference in these

A space image of 2005's Hurricane Katrina shows a well-defined, powerful storm system heading for the Gulf Coast. (*NASA*)

speeds, the resulting wind sheer will cause the hurricane to become unstable, with clouds and winds opposing each other rather than working together in a gigantic swirl that increases in speed. Hurricanes do not tend to generate close to the equator (within five degrees latitude) because the Coriolis effect is stronger father away from the equator, and also because they need a low pressure area that is not close to the equator.

What is a **tropical storm**?

You can think of a tropical storm as a less-intense hurricane (you can also call a hurricane a more intense tropical storm). If wind speeds are between 38 and 74 miles (61 and 119 kilometers) per hour, then it is considered a tropical storm and not a hurricane.

What is a **tropical depression**?

A tropical depression consists of a line of organized storms with wind speeds below 38 miles (61 kilometers) per hour. Tropical depressions have the potential to become tropical storms or hurricanes.

What is the **Coriolis effect**?

Named after French mathematician scientist Gustave Coriolis (1792-1843), who first explained it in 1835, the Coriolis effect refers to the way objects appear to move in a curving or circular pattern when observed from a point of view position that is rotating. Imagine yourself standing next to a playground carousel. Two of your friends are riding on opposite sides of the carousel as it spins around. One friend holds a ball and tries to roll it to the person on the other side, but as he does so, the ball seems to veer to one side and roll off the carousel. To your point of view (as you stand off to the side), however, the ball rolled in a straight line, but it did not reach your other friend because as the ball moved across the carousel moved beneath it and the intended receiver was no longer in the original position.

Now imagine the Earth as it spins on its access. Above the Earth, suspended in the atmosphere, is a forming hurricane. The air around the hurricane is moving toward the eye, which is where the lowest air pressure is. However, as the air moves toward the eye, it is deflected to the right (in the Northern Hemisphere) by the

Does the Coriolis effect make the water in my toilet, sink, and bathtub swirl clockwise?

No, the Coriolis effect has very little effect on such small bodies of water. The flow down the drain is mostly a function of the shape of the container. Interestingly, if your body were completely symmetrical (and no one's is) and neither leg were longer and you were walking on perfectly flat land then you might start veering due to the Coriolis effect.

Earth's spin, or to the left (in the Southern Hemisphere). This causes the hurricane clouds to rotate counterclockwise in the North and clockwise in the South.

What part of a **hurricane** is **most damaging**?

Floods caused by hurricane storm surges are the most destructive element. The low-pressure center of a hurricane causes a mound of water to rise above the surrounding water. This hill of water is pushed by the hurricane's fierce winds and low pressure onto the land, where it floods coastal communities, causing significant damage. Hurricanes sometimes spark tornadoes that contribute to the devastation.

How **fast** do **hurricanes travel**?

A typical hurricane will travel across the ocean at a speed of about 250 miles (400 kilometers) per day, or about 10 to 15 miles (16 to 24 kilometers) per hour. They have been known, though, to advance at speeds as fast as 60 miles (96.5 kilometers) per hour, which was the case during the New England hurricane of 1938.

What is a **storm surge**?

Not to be confused with a tsunami, a storm surge is a sudden upwelling of ocean water caused by winds and pressure changes affecting the water's surface. Hurricanes generate large waves—swells—that radiate outwards in all directions as they travel over the ocean. The swells, which can move toward the shoreline about three or four times faster than the actual storm, arrive on land before hurricanes strike. Before advanced weather systems and the use of satellites, these swells warned people that a hurricane was approaching. The swells become storm surges by the time the main storm arrives, raising water levels as much as 25 feet (7.5 meters) and causing massive coastal flooding. By some estimates, the storm surge resulting from 2005's Hurricane Katrina was 28 feet (8.5 meters).

Why don't we see **hurricanes** in the **South Atlantic Ocean**?

The cold sea surface temperatures of the South Atlantic and atmospheric conditions, such as the tendency of the Intertropical Convergence Zone to remain in the

Northern Hemisphere, make hurricane formation south of the equator unlikely. However, in March 2004, a hurricane did strike the coast of Brazil, which was a very unusual event.

Is a **polar low** like an **Arctic hurricane**?

Some strong hurricanes—such as 1992's Hurricane Andrew—have continued to be active while traveling as far north as the Arctic, but at that point they are no longer considered tropical storms or hurricanes. There is also something called a "polar low," which is like a small hurricane that can form above the Arctic Circle. Polar lows (extra-tropical lows) tend to range from 50 to 250 miles (100 to 500 kilometers) in diameter, versus tropical hurricanes that are easily twice as big in many cases. Not only are they smaller, but polar lows tend to have a shorter lifespan than southern hurricanes, rarely lasting more than 36 hours and more typically only about 12 hours. However, they can still be very intense, generating strong winds and heavy snowfalls.

Has there ever been a **hurricane** in **Great Britain**?

Well, what really occurred in Great Britain was a very intense, low-pressure system with hurricane-force winds. On January 25, 1990, a storm with winds up to 120 miles (193 kilometers) per hour hit Great Britain, killing 45 people and causing over one billion dollars in damage. Even so, Brits often remember the Great Storm of 1987 even less fondly. Though it cannot be classified as a hurricane (hurricanes are tropical events only), it killed 18 people and was, at that time, the worst storm to hit the British Isles in the last three centuries.

Has a **hurricane** ever made landfall in **southern California**?

While tropical storms have, rarely, reached southern California, there is no record of a hurricane ever reaching the coastline there. A deadly tropical storm took 45 lives in 1939, and Tropical Storm Kathleen caused lots of flooding on September 10, 1976. As for the future, who knows? Hurricanes seem to be increasing in size and number in the twenty-first century, and it is possible that one could travel across, say, northern Mexico and then reach southern California.

What is the **Fujiwhara effect**?

Named after Japanese meteorologist Sakuhei Fujiwhara (1884–1950), the Fujiwhara effect is what happens when two hurricanes come close enough to each other that they begin to rotate around a common central point. For this to occur, the two storms generally need to come to within 300 to 900 miles (500 to 1,500 kilometers) of each other; they also need to be of about equal strength to remain in this partnered dance, or else the stronger storm tends to swallow up the smaller storm.

How **fast** do the strongest **hurricane winds** blow?

The strongest hurricanes have winds that reach speeds over 200 miles (322 kilometers) per hour. Friction with the Earth's surface prevents winds from blowing faster than 225 miles (362 kilometers) per hour.

What is the **Saffir-Simpson hurricane scale**?

The Saffir-Simpson Hurricane Damage-Potential scale, which is the full name, is a five-point scale invented in 1971 by engineer Herbert Saffir (1917-2007) and Robert Simpson (1912–), a hurricane expert. Rating hurricanes on a scale of 1 to 5, with 1 being the weakest and 5 the strongest, the scale ranks these storms according to peak wind speeds and the amount of damage they cause.

Saffir-Simpson Hurricane Scale

Force	Wind Speeds mph/kph	Damage
1	74–95 / 137–176	Little, if any, damage to buildings; mobile homes may be damaged, as well as trees and shrubs; some coastal flooding and minor damage to piers.
2	96–110 / 177–204	Some damage to windows, roofs, and doors; more severe damage to mobile homes, piers, and plants; small watercraft break moorings if they are in unprotected areas; low-lying areas flood 2-4 hours before the hurricane arrives.
3	111–130 / 205–241	Mobile homes are destroyed and small residences and utility buildings are damaged; flooding is more pronounced with land lower than five feet above sea level flooded as much as six miles inland.
4	131–155 / 242–287	Buildings see structural failures and roofs may be completely ripped off; lower floors of buildings near the shore are severely damaged; land lower than 10 feet above sea level floods six miles inland; significant erosion of beaches and shoreline.
5	>155 / >287	Roofs of residential and industrial structures crumble; some buildings completely destroyed, and lower levels of most other structures within 500 yards (475 meters) of the shoreline are severely damaged and flooded to up to 15 feet (5 meters) above ground; massive evacuation of residence within 10 miles (18.5 kilometers) of the shore.

How do **hurricanes** get **named**?

The United States introduced the naming system in 1950 in which each hurricane is given a name in alphabetical order. in 1953, the naming convention was taken over by the World Meteorological Organization (WMO), which selected names from library sources and finalized lists during international meetings. Until 1978, all the names were female, but this practice ended with the 1979 hurricane season, and now names alternate between male and female. The names are all either English, Spanish, or French in origin.

The names are chosen to reflect the cultures and languages found in the Atlantic, Caribbean, and Hawaiian regions. When a tropical storm with rotary action and wind speeds above 39 miles (63 kilometers) per hour develops, the

A home owner frets over his house as strong winds from Hurricane Wilma blow a Florida neighborhood in 2005.

National Hurricane Center near Miami, Florida, selects a name from one of the six listings for Region 4 (Atlantic and Caribbean area). Letters Q, U, X, Y, and Z are not included because of the scarcity of names beginning with those letters.

What will the **names** of hurricanes be **through 2013**?

Hurricane Names: 2009 through 2013

2009	2010	2011	2012	2013
Ana	Alex	Arlene	Alberto	Andrea
Bill	Bonnie	Bret	Beryl	Barry
Claudette	Colin	Cindy	Chris	Chantal
Danny	Danielle	Don	Debby	Dorian
Erika	Earl	Emily	Ernesto	Erin
Fred	Fiona	Franklin	Florence	Fernand
Grace	Gaston	Gert	Gordon	Gabrielle
Henri	Hermine	Harvey	Helene	Humberto
Ida	Igor	Irene	Isaac	Ingrid
Joaquin	Julia	Jose	Joyce	Jerry
Kate	Karl	Katia	Kirk	Karen
Larry	Lisa	Lee	Leslie	Lorenzo
Mindy	Matthew	Maria	Michael	Melissa
Nicholas	Nicole	Nate	Nadine	Nestor
Odette	Otto	Ophelia	Oscar	Olga

What was the greatest natural disaster in United States history?

The greatest natural disaster occurred when a hurricane struck Galveston, Texas, on September 8, 1900, and killed 8,000 to 12,000 people. However, the costliest national disaster to date was Hurricane Katrina, which hit the Gulf Coast in August 2005, killing 1,800 people and causing $100 billion in damages.

2009	2010	2011	2012	2013
Peter	Paula	Philippe	Patty	Pablo
Rose	Richard	Rina	Rafael	Rebekah
Sam	Shary	Sean	Sandy	Sebastien
Teresa	Thomas	Tammy	Tony	Tanya
Victor	Virginie	Vince	Valerie	Van
Wanda	Walter	Whitney	William	Wendy

What **hurricane names** have been **retired**?

Hurricane names are retired when a particularly bad storm has caused a great deal of damage and loss of life. Countries affected by such hurricanes will petition the WMO to have the name retired and replaced with a new name.

Agnes (1972)
Alicia (1983)
Allen (1980)
Allison (2001)
Andrew (1992)
Anita (1977)
Audrey (1957)
Betsy (1965)
Beulah (1967)
Bob (1991)
Camille (1969)
Carla (1961)
Carmen (1974)
Carol (1954)
Celia (1970)
Cesar (1996)
Charley (2004)
Cleo (1964)
Connie (1955)
David (1979)
Dean (2007)
Dennis (2005)
Diana (1990)
Diane (1955)
Donna (1960)
Dora (1964)
Edna (1968)
Elena (1985)
Eloise (1975)
Fabian (2003)
Felix (2007)
Fifi (1974)
Flora (1963)
Floyd (1999)
Fran (1996)
Frances (2004)
Frederic (1979)
Georges (1998)
Gilbert (1988)
Gloria (1985)
Hattie (1961)
Hazel (1954)
Hilda (1964)
Hortense (1996)
Hugo (1989)
Inez (1966)
Ione (1955)
Iris (2001)
Isabel (2003)
Isidore (2002)
Ivan (2004)
Janet (1955)
Jeanne (2004)
Joan (1988)
Juan (2003)
Katrina (2005)
Keith (2000)
Klaus (1990)
Lenny (1999)
Lili (2002)
Luis (1995)
Marilyn (1995)
Michelle (2001)
Mitch (1998)
Noel (2007)
Opal (1995)
Rita (2005)
Roxanne (1995)
Stan (2005)
Wilma (2005)

Gustav and Ike, both 2008 hurricanes, may also be retired.

What incident do many literary scholars believe inspired William Shakespeare to write *The Tempest*?

The Tempest is considered to be the last play completed by the famous British playwright William Shakespeare (1564–1616). It is a mystical, romantic play about shipwrecked sailors who find themselves on an island inhabited by good and evil creatures. Many scholars believe it was inspired by the story of the *Sea Venture,* a ship that sank near the Bahamas in 1609 because of a hurricane. The captain's decision to crash the ship on a reef resulted in his saving 150 of his crew.

How many **category 5 hurricanes** have there been since 1920?

The Saffir-Simpson Scale was not invented until 1971; however, reliable records on hurricane wind speeds and storm surges have been kept since the 1920s. According to such data, there have been 31 category 5 hurricanes in the Atlantic since 1928. Eight of those have occurred since 2003 (four in 2005 alone), which is one piece of evidence some climatologists point to when talking about global warming and its effects on tropical storm intensity. Of all those hurricanes, only four have actually made landfall on U.S. territory. Others have struck Central America or islands in the Caribbean, or they have weakened to category 4 or below before reaching Puerto Rico or the Gulf Coast or Atlantic seaboard. Below is a chronological list of category 5 hurricanes.

Category 5 Hurricanes since the 1920s

Hurricane	Year	Areas Affected
"Okeechobee"	1928	PR, FL, GA, Eastern U.S. Seaboard, U.S. Virgin Islands, Bahamas, Guadeloupe, Virgin Islands, Lesser Antilles
"Bahamas"	1932	Bahamas, Northeastern U.S., Newfoundland, Iceland
"Labor Day"	1935	FL, GA, SC, NC, VA, Bahamas, Big Bend
"New England"	1938	Bahamas, NY, RI, MA, CT, NH, VT, Québec
"Fort Lauderdale"	1947	Bahamas, FL, LA, MS
Dog	1950	Eastern U.S. Seaboard, Lesser Antilles
Easy	1951	Bermuda
Janet	1955	Belize, Mexico, Leeward Islands
Cleo	1958	Remained at sea
Donna	1960	PR, U.S. East Coast, Haiti, Cuba, Bahamas, Dominican Republic, Leeward islands, eastern Canada
Ethel	1960	LA, MS
Carla	1961	TX, Yucatán Peninsula, Central U.S.
Hattie	1961	Belize

Hurricane	Year	Areas Affected
Beulah	1967	TX, Yucatán Peninsula, northern Mexico, Greater Antilles
Camille	1969	AL, MS, LA, Cuba, central East U.S. Coast
Edith	1971	TX, LA, Central America, Venezuela, Mexico
Anita	1977	Northern Mexico
David	1979	PR, FL, GA, U.S. and Canadian East Coast, Cuba, Bahamas, Haiti, Dominican Republic, Windward Islands
Alan	1980	TX, Yucatán Peninsula and northern Mexico, Haiti, Jamaica, Windward Islands
Gilbert	1988	TX, south-central U.S., Yucatán Peninsula, Venezuela, Haiti, Dominican Republic, Jamaica
Hugo	1989	Eastern U.S. Coast, PR, U.S. and British Virgin Islands, Dominica, Montserrat, Guadeloupe
Andrew	1992	FL, LA, Bahamas
Mitch	1998	FL, Yucatán Peninsula, Central America
Isabel	2003	Eastern U.S. Seaboard and as far inland as PA and OH, Ontario, Bahamas, Greater Antilles
Ivan	2004	FL, AL, LA, TX, Eastern Seaboard, Cuba, Jamaica, Venezuela, Windward Islands, Grand Cayman
Emily	2005	TX, Yucatán Peninsula and northeastern Mexico, Jamaica, Cayman Islands, Honduras, Windward Islands
Katrina	2005	FL, LA, MS, AL, Eastern Seaboard, Bahamas, Cuba
Rita	2005	FL, AR, MS, LA, TX, Cuba
Wilma	2005	FL, Yucatán Peninsula, Central America, Jamaica, Cayman Islands, Bahamas, eastern Canada
Dean	2007	PR, Jamaica, Haiti, Belize, Nicaragua, Honduras, Mexico, St. Lucia, Dominica, Martinique, Leeward Islands
Felix	2007	Central America, Venezuela, southern Windward Islands

What was the **2008 hurricane season** like?

After a two-year lull in 2006 and 2007, 2008 came roaring back as one of the strongest hurricane seasons on record. Although there were no Category 5 hurricanes, there were two Category 4s (Ike and Gustav) and three Category 3s (Bertha, Omar, and Paloma). Neither Ike nor Gustav made landfall in the United States as Category 5s, but a weakened Hurricane Gustav caused tremendous destruction in Galveston, Texas. About 1,000 deaths resulted from eight hurricanes, five of which were classified as major (Category 3 or 4). There were also eight strong tropical storms that did not develop into hurricanes. Despite lacking a category five storm, the 2008 season broke the record for most months with a major hurricane, with five months including such major storms: July (Bertha), August (Gustav), September (Ike), October (Omar), and November (Paloma). Of these hurricanes, Gustav was the worst for Americans. Galveston had to be evacuated because of a huge storm surge and waves reaching an incredible 50 feet (15 meters); many people lost their homes.

What have been the **costliest hurricanes** in the **United States**?

Top 10 Costliest Hurricanes in the U.S.

Hurricane	Year	Category	Damage (billions)
Katrina (LA nearby states)	2005	3	$100-$135
Andrew (southeast FL & LA)	1992	5	$36
Charley (FL)	2004	4	$14
Hugo (Atlantic seaboard, Charleston, SC)	1989	4	$14
Ivan (AL, FL)	2004	3	$14
Agnes (Atlantic seaboard, FL)	1972	1	$11
Betsy (southeast FL, LA)	1965	4	$10.8
Francis (FL)	2004	2	$8.9
Camille (AL, LA, MS)	1969	5	$7.5
Jeanne (eastern TX)	2004	3	$6.9

What have been the **deadliest tropical storms** since 1900 in the **world**?

Deadliest Hurricanes and Cyclones in the World, 1900 to 2008

Name	Location	Year	Total Deaths
Bhola Cyclone	Bangladesh	1970	150,000-500,000
Unnamed	Bangladesh	1991	131,000-138,000
Cyclone Nargis	Myanmar	2008	100,000-140,000
Two unnamed cyclones	Bangladesh	1965	about 60,000
Unnamed	Bangladesh/India	1963	22,000
Andhra Pradesh Cyclone	India	1977	10,000-20,000
Hurricane Mitch	Central America	1998	11,000-18,000
Orissa Cyclone	India	1971	about 10,000
Cyclone 05B	India	1999	about 10,000
Galveston Hurricane	Galveston, TX	1900	8,000-12,000

What **U.S. cities** have been **wiped out by hurricanes**?

Since the nineteenth century, several U.S. cities have been destroyed, or nearly destroyed, by hurricanes.

U.S. Cities Destroyed by Hurricanes

City	Year
Tampa, FL	1848
Indianola, TX	1886
Galveston, TX	1900
Miami, FL	1926
Pass Christian, MS	1969
Biloxi, MS	1969
Homestead, FL	1992
New Orleans, LA	2005

A mangrove swamp on Yap Island in Micronesia helps prevent shoreline damage from tsunamis and cyclones. Part of the reason many populated areas of the world have become vulnerable is because plant life such as mangrove trees has been removed. (*photo by Ben Mieremet, Senior Advisor OSD, NOAA*)

Why have **cyclones** striking out of the **Indian Ocean** proven to be so deadly?

Cyclones that have struck India, Bangladesh, Myanmar, and other areas in Southeast Asia are no stronger than the powerful hurricanes that have caused so much death and damage in the Gulf Coast. Many more people have died as a result of these cyclones, however, because of high population densities near the coasts and because so many of these people live in poverty and survive in rickety shacks and other substandard shelters. The 2008 cyclone that hit Myanmar was particularly brutal in the area of the Irrawaddy Delta, a low-lying peninsula that had been converted to rice and shrimp farms. These farms replaced mangrove forests that would have served as natural buffers between the land and the oncoming cyclone.

What is the **difference** between a **cyclone** and a **typhoon**?

The words "cyclone" and "typhoon" are simply other words for hurricane. When hurricanes occur west of the International Dateline in the northwest Pacific Ocean, they are referred to as typhoons, and when they occur in the Indian Ocean and near Australia, they are called cyclones.

What is an **anticyclone**?

Rather than a having a low-pressure mass surrounded by swirling air, an anticyclone is an air mass centered around a high-pressure area with air rotating around

A P-3 airplane operated by NOAA flies near the eye of a hurricane (NOAA).

it in a clockwise manner (in the Northern Hemisphere) and counterclockwise south of the equator. As with a cyclone, in an anticyclone air rotates faster the greater the difference between the air pressure at the center compared to the surrounding air.

What is the **eye of a hurricane**?

The eye of a hurricane is a region of relative calm in the middle of the swirling storm. Hurricane eyes range in size from 4 to 40 miles (7 to 74 kilometers) in diameter (although Hurricane Wilma had an eye only 2 miles [3.2 kilometers] wide) and can be so free of clouds that sunshine can be seen. The more intense the hurricane, the smaller the eye tends to be. The eye is surrounded by the "eye wall," an apt phrase because it is literally a circular wall that can reach 7 miles (11.3 kilometers) into the sky. Once past the eye wall, the hurricane resumes, with winds blowing as fast as 150 miles per hour (278 kilometers per hour) or more.

What are **spiral bands**?

The clouds arranged in curving bands that form the outside of a hurricane rather like those in a spiral galaxy are called spiral bands. They can extend for several hundred miles beyond the eye of the hurricane.

What is a **concentric eye wall**?

Most hurricanes are centered around one eye, but sometimes a secondary eye wall surrounds this eye. The second eye wall surrounds the first, and is thus called a concentric eye wall.

What is a hypercane?

Professor Kerry Emmanuel at the Massachusetts Institute of Technology is an authority on using computer models to simulate hurricane activity. In one extreme scenario, Emmanuel hypothesized what might happen if a large asteroid hit one of Earth's oceans. Of course, there would be monstrous waves and immense heat generated by the impact, but there would be another side effect as well: a hypercane. A hypercane would occur after the impact because deep ocean waters would heat up. The result would be a strange hurricane that would only be about 10 to 20 miles (16 to 32 kilometers) across but would have incredible winds reaching 500 miles (800 kilometers) per hour! Such a hypercane would also be capable of jettisoning moisture upwards of 25 miles (40 kilometers) into the atmosphere.

What is the difference between a **hurricane watch** and a **warning**?

Somewhat like with tornado watches and warnings, a hurricane watch means that conditions are good that a hurricane will form within the next 36 hours or so. A hurricane warning means that a hurricane is expected to make landfall within 24 hours. Warnings and watches are issued by the National Weather Service's National Hurricane Center.

What is the **National Hurricane Center**?

Based at Florida International University near Miami, the National Hurricane Center is part of the National Weather Service. Its mission is to predict and warn of dangerous hurricanes in the Caribbean and Gulf of Mexico. The headquarters building itself is heavily fortified against hurricanes. Having 10-inch-thick walls, roll-down shutters for its windows, and being far enough inland to be safe from storm surges, it is designed to survive 130 mile (210 kilometer) per hour winds so that it can still be operational after almost any hurricane.

How much **energy** can a **hurricane produce**?

An average hurricane can strike with a force equivalent to about 400 hydrogen bombs, each packing 20 megatons of explosive power.

Can a **hurricane be stopped**?

For all practical purposes, no. Proposals have been suggested, such as cloud seeding techniques, but to date science has yet to come up with a solution. During the 1950s, the U.S. federal government launched the Stormfury Project, which was an effort to dump silver iodide crystals near the eyes of hurricanes. The theory was that the seeding would generate a secondary eye in the storm, which would then hamper or even cause the original eye to collapse. Several experiments were conducted

Where in the tropics are you safe from a hurricane?

Hurricanes do not strike within five degrees latitude on either side of the equator. Therefore, if you wish to live in the sunny tropics and have a healthy fear of hurricanes, you may want to investigate real estate deals near the equator.

in 1961, 1963, 1969, and 1971, but while sometimes the results seemed promising, the data was inconclusive. Hurricane Esther, in 1961, appeared to be weakened by as much as 30 percent through seeding, but there was no proof that the storm didn't just weaken all by itself. The government gave up the project in the 1970s, and while private companies have continued some of this research, most meteorologists believe that there is just no practical way to destroy a hurricane. The problem seems to be the fact that, for it to work, cloud seeding requires supercooled water, but hurricane clouds contain insufficient supercooled moisture.

What was the **"Storm that Wouldn't Die"**?

In November 1992 Typhoon Gay lasted for days as it traveled thousands of miles across the Pacific Ocean, eventually making landfall in Alaska, British Columbia, and California. At its peak, the storm boasted winds of 200 miles (322 kilometers) per hour. After making landfall in the United States, it continued across the Great Plains, then regained strength as it reached the East Coast. Here, it became a new storm on December 11 with winds of about 90 miles (145 kilometers) per hour.

What was the **"Great Hurricane of 1938"**?

Possibly the worst storm to ever hit the American Northeast since the arrival of Europeans, the "Great Hurricane of 1938" wreaked havoc throughout New England and down as far as Long Island. Six hundred people died as a result of the storm in what has also been called the "Long Island Express."

What was **Hurricane Katrina**?

Hurricane Katrina was the name given to the devastating hurricane that developed in the Atlantic Ocean, crossed the Gulf of Mexico, and struck New Orleans and many other cities along the southern coast of the United States in late August 2005.

How many **people died** as a result of the subsequent **failure of the levees** and flooding after **Hurricane Katrina** struck?

The accepted final figures state that 1,836 people lost their lives following the landfall of Hurricane Katrina.

Hurricane Katrina caused over $100 billion in damages to the United States in 2005, with New Orleans being particularly hard hit.

Was the **2005 New Orleans disaster caused** by a flood or a hurricane?

The initial cause of the disaster was Hurricane Katrina, which whipped up tides and sea water against a very fragile levee system that protected New Orleans. The city is 49 percent below sea level, and so when the man-made levees broke, flood waters moved in and inundated much of the city.

FLOODS

How much rain does it take to **make a flood**?

The amount varies widely for different areas. In some U.S. western deserts, or in some large urban areas, just a few minutes of strong rain will cause a flash flood in canyons and low-lying areas. In areas prone to greater rainfall amounts, it often takes quite a bit more rain (sometimes a few days' or weeks' worth) to cause rivers to overflow and dams to fill up, raising concerns of those who live downstream. Areas that normally receive more rainfall have better natural drainage systems and are usually home to plants that readily absorb the extra water.

What **causes a flood**?

Flooding results when more water enters an environment than can be easily absorbed into the soil or drained away in rivers and streams. Flooding is usually caused by intense rainfalls that dump many inches of water onto an area over a short period of time, or they can also be caused by ocean swells and storm surges

Severe floods are a regular threat to many people living near rivers, oceans, and lakes.

initiated by hurricanes and tropical storms. Tsunamis, naturally, also cause flooding. The 2004 tsunamis that resulted in an undersea earthquake in the Indian Ocean, for instance, killed about 238,000 people in 11 surrounding countries. Most of these people died from the initial landfall of the waves and resulting floods. In addition, floods can be caused artificially, such as when a dam or levee breaks.

What is a **flash flood**?

Floods can happen relatively gradually, such as when water slowly rises over the banks of a river or lake, or suddenly, such as when a dam or levee is damaged. When a flood happens quickly, it's called a flash flood.

What caused all the flooding in **New Orleans** during **Hurricane Katrina**?

Most people now agree that the destruction of New Orleans could have largely been avoided had the canal levees, built by the U.S. Army Corps of Engineers, been up to the task of controlling the storm surges. Because many of the levees failed, 80 percent of the city was engulfed in water.

Why do **flash floods kill** so many people?

Simply put, people don't seem to recognize the power of flash floods and the fact that even a half foot of rushing water can knock down an adult and sweep him or her away. Moving flood waters just a few feet deep are capable of pushing cars and small trucks. Floods can either drown you, or they can kill you by carrying deadly debris. The other factor, of course, is that flash floods catch people by surprise, and

What do many scholars believe caused the flood that Noah and his family survived in the Bible?

Some archeologists believe that, sometime between the years 5400 B.C.E. and 5200 B.C.E., the Euphrates River flooded the surrounding valley, covering an area of about 40,000 square miles (about 104,000 square kilometers). Although this flood did not encompass the globe, to those living in the area it certainly would have seemed like the end of the world and it could have inspired the well-known biblical story.

if they happen at night, those who are sleeping can certainly be caught unawares. Sometimes, too, people can just lack a little common sense. There have been many stories of people who were standing in dry river beds when a flash flood suddenly barreled down on them; instead of running up the river bank and to safety, they ran downriver in the vane hope of outrunning the rushing waters. This is how a number of deaths occurred during the 1976 flood that hit Big Thompson, Colorado.

What are some of the **worst floods in history** caused by inclement weather?

Not all lethal floods in recorded history have been caused by bad weather. For example, seawall failures in the Netherlands have, on several occasions, caused tragic floods that killed thousands. The table below, though, lists some of the worst weather-related floods in history.

Worst Weather-related Floods and Death Tolls

Location	Death Toll	Year
Yellow River, China	1 to 3.7 million	1931
Yellow River, China	1 to 2 million	1887
Yangtze River, China	145,000	1931
England and the Netherlands	100,000	1099
Caracas, Venezuela	10,000	1999

What was the **Great Flood of 1993**?

Heavy rainfall caused river flooding in 1993 that was so severe in Iowa that, when observed by NOAA sensors, the state looked like a sixth Great Lake. The Mississippi River was seven miles wide at some points, and the Missouri River also spilled over its banks, causing nearly $20 billion in property losses, 48 deaths, and the evacuation of 85,000 people.

What have been some of the **most destructive floods** in history?

In the United States, the failure of a dam in 1889 upstream from the community of Johnstown, Pennsylvania, killed 2,200 people. Some of the world's most catastrophic flooding takes place in China. A flood on the Huang He River in 1931 killed 3.1 million people.

What is a **floodplain**?

A floodplain is the area surrounding a river that, when unmodified by human structures, would normally be flooded during a river flood. A floodplain can be a few feet or many miles wide, depending on the river flow as well as the local terrain. Even though levees and flood walls can be built (with homes and businesses built just behind them), the floodplain does not vanish. If the structures break or are damaged, the water from a flood can fill a floodplain, just as it did before humans occupied it.

Why do **people live** in **floodplains**?

People have lived in floodplains for thousands of years. Fertile land for agriculture lines the floodplain, and the nearby water source makes life easier. Unfortunately, when the river does flood, these communities are severely damaged and people suffer. Hazard mitigation, such as levees, dams, dikes and other structures, attempt to limit damage during floods. Sometimes, when the structures fail (such as a levee breaking), large areas are inundated with water. Inhabitants of floodplains must balance the risks with the rewards of living in such an unpredictable environment.

What is a **100-year flood**?

A 100-year flood refers not only to the size of a flood, but also to the odds of it occurring. A 100-year flood has a one percent (or 1 in 100) chance of occurring in any given year. It has no relationship to the frequency of occurrence. The magnitude of such a flood is relative to the frequency of occurrence, so a 100-year flood is much larger than any run-of-the-mill annual flood. A 500-year flood has only a one in 500 (0.2 percent) chance of occurring in any given year and would be much larger and more devastating than a 100-year flood.

What is the **National Flood Insurance Program**?

The National Flood Insurance Program (NFIP) was established by the U.S. federal government in 1956 as a subsidized insurance program for home and business owners. The government began the program by creating Flood Insurance Rate Maps (FIRM) showing the boundaries of 100-year and 500-year flood zones. The cost of the insurance is based on the flood risk. The Federal Emergency Management Agency (FEMA) oversees the program and requires the purchase of flood insurance by any owner affected by a disaster before they can be provided with disaster assistance. This way, the next time a flood occurs, they will be insured.

How can I obtain a **flood map** of my community?

The best way to see a Flood Insurance Rate Map (FIRM) for your area would be to contact your local government. Their planning or emergency management agency should have the FIRM maps available. Purchasing them from FEMA is not recom-

A twister strikes Union City, Oklahoma, on May 24, 1973. Oklahoma is probably the most dangerous U.S. state to live in when it comes to tornadoes. *(NOAA Photo Library, NOAA Central Library; OAR/ERL/National Severe Storms Laboratory)*

mended because the maps change often and are best interpreted by a planning or emergency expert.

What should I do in the event of a **flood**?

If a flood is expected, turn on your battery-powered radio and listen for information about when and where to evacuate. If a flood or flash flood is coming toward you, move quickly to a higher elevation—but don't ever try to outrun a flood. Also, don't drive through standing water, as it can quickly rise and stall your vehicle, possibly trapping you among swift water.

TORNADOES

Which is **more deadly: tornadoes or lightning** ?

More people in the United States die from being struck by lightning than from a tornado. About 80 people die annually from tornadoes, while more than 100 perish after a lightning stroke, on average.

What are **tornadoes**?

Tornadoes are very powerful, yet tiny storms that have destructive winds capable of leveling buildings and other structures. Winds in a tornado form a dark gray column of air (though white, bluish, or even red is possible, depending on how the Sun's light rays might be reflecting off the tornado), with the center of the tornado acting like a vacuum, picking up objects and moving them along the storm's path. Tornadoes can last from a few minutes to an hour or more. They are one of the most destructive forces in nature, with the largest tornadoes sometimes having diameters of over one mile (1.6 kilometers) and wreaking a path of destruction over a hundred miles (160 kilometers) long.

Who was the **first person** to **seriously study tornadoes**?

John P. Finley (1854–1943), an officer in the U.S. Army Signal Service and meteorologist, was the first to write an authoritative book on the subject: *Tornadoes,* which was published in 1887.

Are all **funnel clouds tornadoes**?

No, in order to be classified as a tornado, the vortex needs to touch both a cloud above and the ground below. If it does not, then it is just considered to be a funnel cloud. Also, not all tornadoes have visible funnel clouds. If there is not dust and debris in the vortex, the tornado may be invisible or nearly invisible to the naked eye. Interestingly, you *can* have a tornado *without* a funnel cloud, because a cloud that descends to the ground without rotation can also be categorized as a tornado.

Supercells, such as this one seen in Miami,Texas, in 1980, produce powerful storms and tornadoes. (*NOAA Photo Library, NOAA Central Library; OAR/ERL/National Severe Storms Laboratory*)

Where does the **word "tornado"** come from?

The word comes from the Latin root *tornare,* which means to turn. *Tronada,* in Spanish, means thunderstorm and is also considered a source of the word tornado.

What is a **willy-willy**?

Willy-willy is the Australian name for a tornado or dust devil.

What **causes a tornado**?

Scientists still do not fully understand how a tornado forms and why. The general theory is that tornadoes form from cloud systems that are slowly spinning, usually in supercell storms, but weaker systems can produce them as well. The current belief is that a mesocyclone within a storm system is surrounded by pronounced variations in air temperature and downdraft. However, tornadoes have been known to form in cloud systems where strong winds and variations in temperature do not exist. Once a funnel begins to form, however, it can gain speed and strength in much the same way as ice skaters gain rotation speed as they pull their arms close to their bodies.

Are **tornadoes** often formed in **hurricanes**?

It is fairly common for tropical storms and hurricanes to generate tornado activity, and sometimes several tornadoes touch down in a single storm. One of the best examples of this was 1967's Hurricane Beulah, during which 115 tornadoes were seen. In 2004 Hurricane Frances generated 123 tornadoes, the most ever recorded.

What **causes a tornado to dissipate**?

Just as the formation of tornadoes is not well understood, neither are the reason for why they disappear. One theory is that when colder air begins to flow out of the storm—the mesocyclone at the center of the storm becomes surrounded by cooler air—the tornado is robbed of sufficient energy and loses power. However, this is not a hard and fast rule, and cool air outflows have been observed that actually are followed by tornado formation.

About how **many tornadoes** are recorded in the United States **annually**?

A typical tornado season sees about 800 tornadoes in the United States. However, better observational techniques, including advanced Doppler radar and growing numbers of tornado chasers, has led to increases in tornado observations since around 1990, so this official average might go up.

What is the **Fujita and Pearson Tornado Scale**?

The Fujita and Pearson Tornado Scale—usually just referred to as the Fujita Scale—was introduced in 1971 by University of Chicago professor T. Theodore Fujita (1920–1998) and Allen Pearson (1925–), who was then the director of the National Severe Storms Forecast Center. The scale ranked tornadoes by their wind speed, path, length, and width. The ranking ranges from F0 (very weak) to F5 (incredibly destructive). This scale was replaced in 2007 by the Enhanced Fujita Scale.

Fujita and Pearson Tornado Scale

Scale	Speed (mph/kph)	Damages
F0	40–72/64–116	Light damage: damage to trees, billboards, and chimneys
F1	73–112/117–180	Moderate damage: mobile homes pushed off their foundations and cars pushed off roads
F2	113–157/181–253	Considerable damage: roofs torn off, mobile homes demolished, and large trees uprooted
F3	158–206/254–331	Severe damage: even well-constructed homes torn apart, trees uprooted, and cars lifted off the ground

Scale	Speed (mph/kph)	Damages
F4	207–260/332–418	Devastating damage: houses leveled, cars thrown, and objects become flying missiles
F5	261–318/419–512	Incredible damage: structures lifted off foundations and carried away; cars become missiles. Less than 2% of tornadoes are in this category
F6	319–380/513–611	No F6 has been recorded, but if such a twister occurred it would be absolutely devastating

What is the **Enhanced Fujita Scale**?

Proposed by the National Weather Service in February 2006, and first put into use on February 1, 2007, the Enhanced Fujita Scale (EFS) was created to better reflect actual damages recorded since the original Fujita Scale was developed. Meteorologists have recently concluded that structures could be damaged by tornadic winds that were slower than previously thought. The original scale, which was felt to be too general, did not take into careful enough account the different types of construction, and it was hard to evaluate tornadoes that struck in low-populated areas where not many structures were present. The new scale also offers more detailed descriptions of potential damages by using 28 Damage Indicators that describe building types, structures, and vegetation, accompanied by a Degrees of Damage scale. Otherwise, the EFS uses the same categories, ranking tornadoes from 0 up to 5.

Enhanced Fujita Scale

Scale	Wind Speed (mph/kph)	Damages
EF0	65–85/105–137	Tree branches break off, trees with shallow roots fall over; house siding and gutters damaged; some roof shingles peel off or other minor roof damage.
EF1	86–110/137–177	Mobile homes overturned; doors, windows, and glass broken; severe damage to roofs.
EF2	111–135/178–217	Large tree trunks split and big trees fall over; mobile homes destroyed, and homes on foundations are shifted; cars lifted off the ground; roofs torn off; some lighter objects thrown at the speed of missiles.
EF3	136–165/218–265	Trees broken and debarked; mobile homes completely destroyed and houses on foundations lose stories, and buildings with weaker foundations are lifted and blown distances; commercial buildings such as shopping malls are severely damaged; heavy cars are thrown and trains are tipped over.
EF4	166–200/266–322	Frame houses leveled; cars thrown long distances; larger objects become dangerous projectiles.
EF5	>200/>323	Homes are completely destroyed and even steel-reinforced buildings are severely damaged; objects the size of cars are thrown distances of 300 feet (90 meters) or more. Total devastation.

Are there any other tornado strength scales in use?

Yes. In Great Britain meteorologists use the TORRO (the TORnado and storm Research Organisation) Scale, which is divided into T0 through T10 measurements of strength. A T0 tornado has winds beginning at 40 miles (64 kilometers) per hour, and a T10 is a "Super Tornado" with wind speeds of 270 to 299 miles (435 to 480 kilometers) per hour.

How **strong** does a tornado have to be to be considered **"significant"**?

Tornadoes that are EF2 or stronger are generally considered to be significant in scale. EF4 and EF5 tornadoes are considered "violent," and rightly so!

How many **F5 tornadoes** have there been in the **United States since 1970**?

There have been 28 F5 tornadoes in the United States since 1970. Six of these occurred during the tornado outbreak of April 1974.

F5 Tornadoes in the United States, 1970–2008

Location	Date
Lubbock,TX	May 11, 1970
Delhi, LA	February 21, 1971
Valley Mills, TX	May 6, 1973
Daisy Hill, IN	April 3, 1974
Xenia and Sayler Park, OH	April 3, 1974
Brandenburg, KY	April 3, 1974
Mt. Hope, AL	April 3, 1974
Tanner, AL	April 3, 1974
Guin, AL	April 3, 1974
Spiro, OK	March 26, 1976
Brownwood, TX	April 19, 1976
Jordan, IA	June 13, 1976
Birmingham, AL	April 4, 1977
Broken Bow, OK	April 2, 1982
Barneveld, WI	June 7, 1984
Niles, OH	May 31, 1985
Hesston, KS	March 13, 1990
Goessel, KS	March 13, 1990
Plainfield, IL	August 28, 1990
Andover, KS	April 26, 1991
Chandler, MN	June 16, 1992
Oakfield, WI	July 18, 1996
Jarrell, TX	May 27, 1997
Pleasant Grove, AL	April 8, 1998
Waynesboro, TN	April 16, 1998
Bridge Creek & Moore, OK	May 3, 1999
Greensburg, KS	May 4, 2007
Parkersburg, IA	May 25, 2008

What is a turtle?

In terms of meteorology, the word turtle does not refer to a shelled reptile. Rather, they are sturdy encased instruments that can be placed in the path of a tornado to measure air pressure, humidity, and temperature. Unfortunately, they cannot measure wind speeds, but they do gather this other valuable data.

Has there ever been an **F6 tornado**?

No. Although the old Fujita Scale did allow for an F6 tornado (estimating that winds up to 380 miles [611 kilometers] per hour were theoretically possible), there has been no recorded tornado of that intensity. The EFS does not include an F6 category at all, and lists any tornado with winds over 200 miles (320 kilometers) per hour as an F5.

Do meteorologists actually have **reliable data** on tornado **wind speeds**?

No, and that is a big part of the problem in categorizing tornadoes. Tornado wind speeds have been scientifically estimated using Doppler radar and video observations, but there have been no successful attempts to physically measure speeds using an anemometer.

What is the **VORTEX project**?

VORTEX stands for Verification of the Origin of Rotation in Tornadoes Experiment. Conducted at the National Severe Storms Laboratory from 1994 to 1995, the project was led by Erik Rasmussen, and the goal was to conduct intense data gathering using weather balloons, air planes, turtles, and cars equipped with radars and taking photographs. The study focused on tornadoes formed in supercells and collected so much data that it took years to analyze it all.

How long a **path** do **tornadoes typically make**?

Most tornadoes actually have fairly brief lifespans, lasting less than an hour and often only several minutes or seconds. About five miles of travel can be expected from an average tornado. Of course, there have been much longer-lived tornadoes, such as one on March 18, 1925, which crossed 215 miles (346 kilometers) of terrain.

What is a **multiple-vortex tornado**?

Sometimes a central tornado can be surrounded by smaller tornadoes called subvortices or suction vortices. There can be as many as seven subvortices surrounding the central tornado, though two to five is more common. Interestingly, these smaller tornadoes tend to be more intense, with winds spinning at about 100 miles (160 kilometers) per hour or more faster than the central vortex.

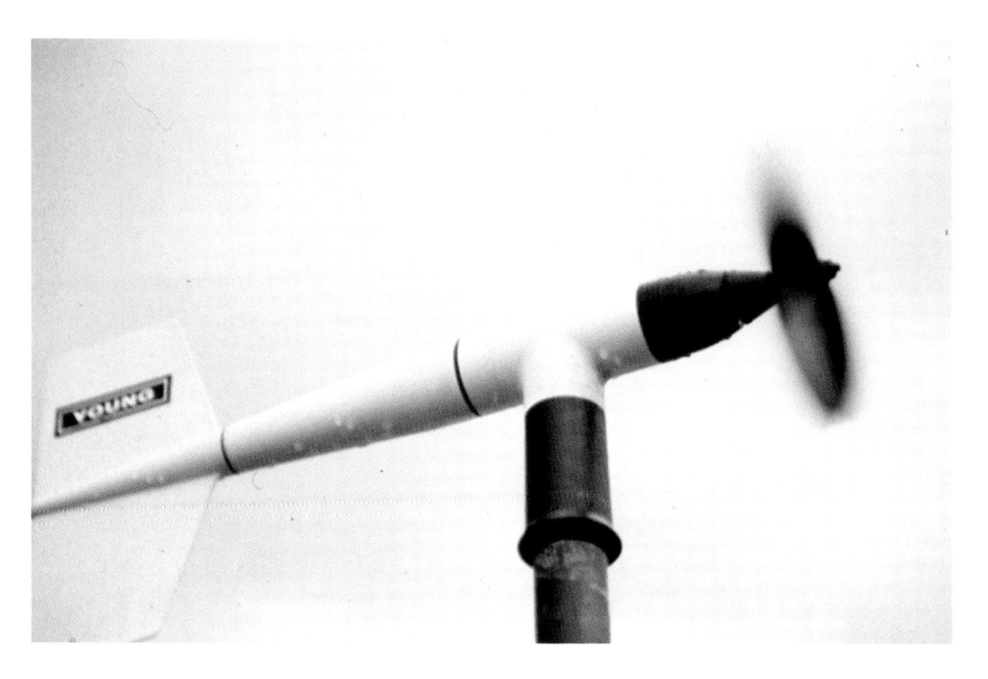

A wind speed instrument mounted on a Project Vortex vehicle monitors conditions in northern Texas in this 1994 photo. *(NOAA Photo Library, NOAA Central Library; OAR/ERL/National Severe Storms Laboratory)*

What is the difference between **subvortices** and **satellite tornadoes**?

The difference is often hard to distinguish, but satellite tornadoes form independently of other, larger tornadoes, while subvortices in multiple-vortex tornadoes form within a central tornado.

What is more dangerous, a **big tornado** with a funnel spanning over a mile in diameter, or a **narrow, speedy tornado**?

Either one can be equally dangerous. Many factors are involved when considering how lethal and destructive a tornado may be, including how fast the vortex is spinning and how much debris is being blown around.

What is a **rope tornado**?

A rope tornado is simply a descriptive phrase referring to a tornado that has a thin, bending, rope-like appearance. Some people make the bad mistake of thinking that rope tornadoes are regular tornadoes that are beginning to dissipate (dissipating tornadoes are described as "roping out.") Actually, they can be just as intense as tornadoes with much wider funnels.

What is a **wedge tornado**?

A wedge tornado is one that appears as a thick, squat column. The observable height of the funnel is about the same as the width.

Powerful tornadic winds wreaked havoc on Union City, Oklahoma, in May 1973. *(NOAA Photo Library, NOAA Central Library; OAR/ERL/National Severe Storms Laboratory)*

What are some of the **erratic behaviors** of **tornadoes**?

Tornadoes are famous for their ability to destroy one home or other building while leaving the structures next door completely intact. Such random acts of destruction occur either when there is one very slim tornado or there is a multiple-vortex tornado or a tornado with several satellite tornadoes surrounding it. A rope tornado, for instance, can clear a path on the ground that is only a matter of several yards wide. On the other hand, the subvortices (or suction vortices) of a multiple-vortex tornado can have widely varying speeds and widths. Sometimes, too, a tornado might not quite reach the ground, but it might be low enough to rip off the roofs of multi-story structures while leaving ranch homes and strip malls unharmed. Tornadoes can exhibit other erratic behaviors, too, weakening and strengthening as they travel, or circling back on themselves to hit the same location twice. The strength of the buildings, too, matters a great deal, as the developers of the Enhanced Fujita Scale recognized.

What is **Tornado Alley**?

The United States has more tornadoes than any other nation, and the majority of these twisters occur throughout the central states, an area known as Tornado Alley. About 200 tornadoes occur in Tornado Alley every year. The area is centered on northern Texas, Nebraska, Kansas, and Oklahoma (which has the highest percentage of severe tornadoes). Tornado activity, however, is also quite high in Louisiana, Mississippi, Alabama, Missouri, Arkansas, Kentucky, Tennessee, Illinois, Indiana, Iowa, and Ohio. The reason why there are so many tornadoes in Tornado Alley is that warm, moist air from the Gulf of Mexico moves northward and hits the dry air coming from the West as well as uplifting air currents. These are the perfect conditions, evidently, for tornado formation.

What is the **most dangerous U.S. state** to live in due to **tornadoes**?

Massachusetts, because of its population density, has actually seen more deaths and severe injuries from tornadoes than any other U.S. state. However, it is still far riskier to live in Oklahoma, which gets hit by more tornadoes than any other state.

What have been the **deadliest tornadoes** on record in the **United States**?

Deadliest U.S. Tornadoes

Location	Date	Deaths
Tri-State (IL, IN, MO)	March 18, 1925	695
Natchez, MS	May 6, 1840	317
St. Louis, MO	May 27, 1896	255
Tupelo, MS	April 5, 1936	216
Gainesville, GA	April 6, 1936	203
Woodward, OK	April 9, 1947	181
Amite, LA/Purvis, MS	April 24, 1908	143
New Richmond, WI	June 12, 1899	117
Flint, MI	June 8, 1953	115
Waco, TX	May 11, 1953	114
Goliad, TX	May 18, 1902	114
Omaha, NE	March 23, 1913	103
Mattoon, IL	May 26, 1917	101
Shinnston, WV	June 23, 1944	100
Marshfield, MO	April 18, 1880	99
Gainesville & Holland, GA	June 1, 1903	98
Poplar Bluff, MO	May 9, 1927	98
Snyder, OK	May 10, 1905	97
Worcester, MA	June 9, 1953	94
Natchez, MS	April 24, 1908	91
Starkville, MS, & Waco, AL	April 20, 1920	88
Lorain & Sandusky, OH	June 28, 1924	85
Udall, KS	May 25, 1955	80
St. Louis, MO	September 29, 1927	79
Louisville, KY	May 27, 1890	76

What is the **average annual number of deaths** in the United States caused by **tornadoes**?

About 60 people die each year in the United States because of tornadoes. Most deaths are caused by flying debris.

Did a tornado almost change music history?

If you noticed the listings of cities in the "Deadliest U.S. Tornadoes" table, you may have caught the fact that the fourth listing was Tupelo, Mississippi, the home town of Elvis Presley (1935–1977). Elvis was just an infant at the time, and he fortunately survived the funnel cloud that killed over 200 people on April 5, 1936.

What was the **deadliest tornado** ever in the **world**?

On April 26, 1989, a tornado landed about 40 miles (65 kilometers) north of Dhaka, Bangladesh. When it was over, 1,300 people were dead, 15,000 were hurt, and about 100,000 lost their homes.

What **months** are the **most likely times for tornadoes**?

The height of the tornado season in the United States is March through August, though tornadoes do form in the South during the off season. More tornadoes occur in May than any other month, but more deaths have been logged during the month of April. One of the worst outbreaks of tornadoes ever recorded ran from April 3 through 4 in 1974. Thirteen states (Illinois, Indiana, Michigan, Ohio, Tennessee, Kentucky, Alabama, Mississippi, West Virginia, Virginia, North Carolina, South Carolina, and Georgia) saw 148 tornadoes touch down, killing 315 people and injuring almost 5,500. The tornadoes cut through 2,500 miles (4,000 kilometers) of terrain.

Besides the **1974 tornado outbreak**, what are some **other serious outbreaks** that have hit the United States?

On April 11, 1965, Palm Sunday, the American Midwest was struck by 37 tornadoes that killed 256 people and injured more than 5,000. An outbreak of 22 tornadoes on May 28, 1984, killed 57 in the Carolinas, left 1,248 injured, and caused about $200 million in damages. (The next year, on May 31, 75 people died after 41 tornadoes raged across Ontario, Canada, causing $450 million in damages). There were two terrible outbreaks in the early 1990s. From April 26 to 27, 1991, 54 tornadoes made landing in Texas and Iowa. Twenty-one people were killed and 208 injured. The next year, on November 21, 94 tornadoes destroyed lives and property through 13 states, stretching from the Midwest to the East Coast. This outbreak left 26 people dead and 641 injured.

Are there **certain times of the day** when **tornadoes are more likely** to happen?

Tornadoes can occur at any hour of the day, but 40 percent of them strike between 2:00 and 6:00 P.M. The danger of a nighttime tornado is that people are often asleep and unprepared for when the warnings are sounded.

A wall cloud descends over Oklahoma in 1984. (*NOAA Photo Library, NOAA Central Library; OAR/ERL/National Severe Storms Laboratory*)

What is a **mesocyclone**?

A mesocyclone is the vortex of air—usually between 1 to 6 miles (2 to 10 kilometers) in diameter—often found within a supercell or other large thunderstorm with cumulonimbus clouds. Wind shear resulting in abrupt changes in wind direction or speed in the storm causes air to circulate in a rolling fashion parallel to the ground; an updraft can then orient the swirling air vertically, thus forming a vortex perpendicular to the ground.

What is a **wall cloud**?

Wall clouds can be warnings that a tornado is about to form in a bank of cumulonimbus clouds. As a mesocyclone expands and gains strength below the cumulonimbus clouds, it begins rotating as warm, humid air moves upward and condenses. Convergence causes the gathering wall cloud to rotate cyclonically, though slowly compared to the tornadoes it could form. Storm chasers who observe such wall clouds know that they can form full-fledged tornadoes within an hour's time.

What is a **beaver tail**?

A beaver tail is a colorful name for a broad, flat, descending cloud that can be seen during some rain storms. Beaver tails usually form outside any areas of falling precipitation, and they tend to swirl inward toward a wall cloud.

Storm chasers, both professional meteorologists and amateurs, provide additional eyes and research observations that have proven helpful to the National Weather Service and NOAA. (*NOAA Photo Library, NOAA Central Library; OAR/ERL/National Severe Storms Laboratory*)

Who are **storm chasers**?

Storm chasers are scientists and amateur storm enthusiasts who track and intercept severe thunderstorms and tornadoes. Two reasons for storm chasing are: 1) to gather data to use in researching severe storms and 2) to provide a visual observation of severe storms indicated on radar stations. In addition, television personnel will chase storms to produce a dramatic storm video. Storm chasing can be an extremely dangerous activity in which strong winds, heavy rain, hail, and lightning threaten one's safety. Individuals who chase storms are trained in the behavior of severe storms.

Roger Jensen (1933–2001) is generally considered the first person to be an active storm chaser. A self-trained weather observer and professional photographer, Jensen spent 50 years recording data on tornadoes as well as thunderstorms. David Hoadley (1938–) is also considered a pioneer in the field and founder of the first newsletter on the subject, *Storm Track*. The first scientist who became a storm chaser was Neil Ward (1914–1972), who worked for the National Severe Storms Laboratory in Norman, Oklahoma, and is considered the official "father of the storm chase" because of his credentials.

How is a **storm chaser** different from a **storm spotter**?

A storm spotter is someone who does not spend his or her time driving around the country looking for dangerous storms. But a spotter is still a vigilant observer who

What is the Skywarn Program?

Organized by the National Weather Service (NWS) in the early 1970s, the Skywarn Program is a concerted effort to organize professional and amateur weather watchers to help provide early warnings to the public about tornadoes and other hazardous weather conditions. You do not have to be a qualified meteorologist to join the Skywarn Program, which hires people who can train you to be a storm spotter. Many people who are amateur radio operators have been particularly enthusiastic volunteers to the Skywarn. Organizations such as the American Relay Radio League (ARRL), the Amateur Radio Emergency Services (ARES), and Radio Amateur Civil Emergency Service (RACES) all work with the NWS, as well as the Federal Emergency Management Agency (FEMA), to help save lives with the Skywarn Program.

tries to find local signs of potentially dangerous storms and report them to the National Weather Service.

What is a **bear's cage**?

The region of heavy rain and hail that usually forms on the southern and western side of a mesocyclone, but to the northeast of where a tornado may exist or may be forming, is referred to by storm chasers as the bear's cage. This is the most dangerous part of the storm, as a tornado can be obscured behind a rain shield, striking with little warning. Visibility is low, and damaging hail, strong winds, and flash floods are also a hazard.

Are there ways **I can anticipate a tornado**?

Tornadoes are hard to predict with any certainty, and even meteorologists can only issue tornado advisories when conditions are right. Some people believe that hail, wind, or lightning will always precede a tornado, but that is not always the case, though large hailstones and other inclement weather do often occur beforehand. Another belief is that if you observe the readings on a barometer suddenly dropping, you can expect a tornado to soon form. This is not an accurate method of prediction, either. Such air pressure drops can occur hours or days before an actual tornado hits. The best strategy is to simply listen closely to weather forecasts during bad storms and trust meteorologists to issue warnings when the conditions merit caution.

Has there ever been a **tornado in Los Angeles**?

Yes, southern California, including Los Angeles, have experienced weak tornadoes on occasion. Fortunately, no deaths have yet been reported in the state as a result. On May 22, 2008, two tornadoes touched ground in Riverside County near San Diego. A tornado warning was also issued in Los Angeles that same month, causing minor damage to homes in the suburb of Inglewood. Los Angeles County has officially seen more than 30 tornadoes since 1918.

A TOtable Tornado Observatory (or TOTO) is placed in the path of oncoming tornadoes to measure air pressure, humidity, temperature, and other information that could prove useful to researchers. *(NOAA Photo Library, NOAA Central Library; OAR/ERL/National Severe Storms Laboratory)*

Do **tornadoes strike** in countries **other than the United States and Canada**?

Yes, but the United States keeps the best records on tornadoes, and so it is difficult to ascertain the frequency and ferocity of twisters in foreign nations. The Canadian prairie experiences significant incidents, but one could say that these are all part and parcel of the North American conditions that are ideal for tornado formation. Other countries that have significant accounts of tornado activity include Great Britain, Italy, western France, Brazil, Argentina, Russia, Bangladesh, China, northern India, Pakistan, Japan, South Africa, and New Zealand. England experiences a tornado about once every year and a half, and Australia probably has more tornadoes than are witnessed, because many of them likely occur in the Outback, where the population is sparse.

Do tornadoes always **turn counterclockwise**?

As a rule of thumb, a tornado in the Northern Hemisphere will rotate counterclockwise, while those in the Southern Hemisphere twist in a clockwise rotation. But, as with any rule, there are always exceptions. Anticyclonic tornadoes (rotating clockwise in the Northern Hemisphere) have occasionally been observed. When they do, they are typically weaker twisters associated with weak storm cells or sometimes appearing as waterspouts. One of the strongest anticyclonic tornadoes was observed in 1998 near Sunnyvale, California. Even rarer—but still possible—is an event when a supercell generates both cyclonic and anticyclonic tornadoes.

What is the difference between a **tornado watch** and a **warning**?

A tornado watch means that weather conditions within the next few hours are favorable for the formation of tornadoes. When a watch is issued, it is wise to listen to radio or television reports for updates, and you should make any preparations necessary in case you need to seek immediate shelter. A tornado warning means that an actual tornado has been spotted on Doppler radar, or by an observer in your area, and you should immediately take shelter.

What should I do when a **tornado** approaches?

Try to get to the lowest level of the building (unless you are in a mobile home or outdoors, in which case you should seek a sturdy and safe shelter; many mobile home parks have a centrally located tornado shelter these days). Go to the center of the room and hide under a sturdy piece of furniture. Stay away from windows, hold on to the leg of a table or something else stable, and protect your head and neck with your arms. If your home has a basement, take shelter there. If not, interior bathrooms are usually the sturdiest rooms in a home, and you can protect yourself further by climbing into the bathtub.

Is it a good idea to **open windows and doors** so that **air pressure is equalized** inside a house?

There is a long-standing myth that when a tornado is near a home it lowers the air pressure outside so much that the higher, interior pressure in a house will cause it to explode. This is not true; tornadoes damage and rip apart houses simply by virtue of wind speeds and the debris they blow about. Opening doors and windows increases the possibility that flying objects will enter the home and possibly hit those hiding inside.

Where is the **best place to hide** from a tornado if I'm **outdoors**?

If there is no other shelter around, storm experts usually recommend that you find a trench or ditch if a tornado is approaching and you don't have time to get to a tornado shelter or basement. (It is *not* a good idea to hide under a freeway overpass, as debris can still easily blow inside and prove deadly.) Some people believe that you are safer if you are in a mountain valley, because the mountains will block tornadoes. Actually, tornadoes have swept through valleys, and they have even touched on mountain peaks. Tornadoes have been recorded in the Teton wilderness area at elevations of more than 10,000 feet (3,000 meters), and during the super tornado outbreak of 1974 many tornadoes struck high areas of the Appalachians.

There is also a myth that standing near a river is a safe place to be during a tornado warning. Again, there have been a number of incidents during which tornadoes have been spotted near, or even crossing, streams and rivers as big as the Mississippi and Missouri. It is not unheard of to learn of boats that have been sunk while sailing down a river when a tornado struck.

Which **month** is the **most dangerous for tornadoes** in the United States?

According to one study, May is the most dangerous month for tornadoes in the United States, with an average of 329, while February's average is the safest with only three. In another study the months December and January were usually the safest, and the months having the greatest number of tornadoes were April, May, and June. In February, tornado frequency begins to increase. February tornadoes tend to occur in the central Gulf states; in March the center of activity moves eastward to the southeastern Atlantic states, where tornado activity peaks in April. In May the center of activity is in the southern Plains states; in June this moves to the northern Plains and Great Lakes area (into western New York). The most costly outbreak of tornadoes occurred in May 1999, when at least 74 tornadoes touched down in less than 48 hours in Oklahoma and Kansas, including an F5 on the outskirts of Oklahoma City causing $1.1 billion in damage.

What is a **gustnado**?

Seen near thunderstorm outflows, a gustnado is a weak vortex that does not touch the clouds. Gustnadoes usually do little more damage than breaking some tree branches and overturning lawn furniture.

What are **waterspouts**?

Often associated with tropical cyclones, waterspouts are tornadoes that can form over a body of water. Instead of sucking up dust and debris, the funnel is visible because of the water it carries. While they tend to be weaker than tornadoes, they can still be lethal and have been known to destroy small boats and cause damage to larger vessels. The most common place to find waterspouts in the United States is off the coast of southern Florida.

What is a **landspout**?

A landspout is, technically, a tornado, albeit a very weak one. Landspouts generally form from non-supercell storms. Despite tending to be less strong than other tornadoes, they have been known to cause fatalities and should still be avoided at all costs.

A waterspout is seen off the coast of the Florida Keys. (*photo by Joseph Golden, NOAA*)

What are **dust devils**?

These columns of brown, dust-filled air, which can rise dozens of feet, are not as evil as the name suggests. They are caused by warm air rising on dry, clear days. Winds associated with dust devils can reach up to 60 miles (96.5 kilometers) per hour and cause some damage. There have been reports of dust devils as tall as 5,000 feet (1,500 meters). They are generally not as destructive as tornadoes and usually die out pretty quickly, though some have been known to displace as much as 50 tons (4,500 kilograms) of dust and light debris. One of the biggest dust devils ever recorded was seen in Bonneville Salt Flats in Utah. It was hundreds of feet tall, and it continued on its way for about 40 miles (65 kilometers).

What is a **steam devil**?

In the Arctic (and, less often, Antarctic, or any other place where conditions are right), cold air passing over warm areas of water can cause steam to rise, and when a whirlwind sweeps in at the same time, this steam or fog forms small steam devils.

Are there **other types** of **tornado-like** whirlwinds?

Certainly. Smoke from forest fires and ash and steam from volcanoes can often stir about in vortices that look like weak tornadoes.

ATMOSPHERIC PHENOMENA

LIGHTNING

What is **lightning**?

Lightning is an electrical discharge occurring in the atmosphere. It comes in many forms and the manner in which it works is only now becoming fully understood.

How does **lightning work**?

The anatomy of a lightning stroke is surprisingly complex. For lightning to occur, of course, there needs to be a source of electricity: an electrical storm. Thunderstorms contain clouds that work like capacitors: the tops of the clouds have a positive charge and the bottoms are negatively charged. Scientists believe that this electrical charge may build up as the result of electrons being exchanged by particles within the clouds as they collide with one another due to temperature differences within the cloud, though just how this happens is a matter of debate. As the difference between the positively charged cloud tops and negatively charged bottoms increases, an electric field is generated.

As the charge within the cloud builds, it has an effect on the surface of the Earth directly below, as well as objects on top of the ground. The negative field on the bottoms of the clouds repels electrons on the ground, which in turn causes the land below the storm to become positively charged. As the charge builds and builds, the air between Earth and cloud becomes ionized—air molecules break down into electrons and positively charged ions—transitioning into a plasma state. This plasma now serves as a conductor between the clouds and the ground below (or to other clouds, or into the surrounding air).

Next, the cloud sends out "stepped leaders," which are precursors to the oncoming flash of lightning. Think of them as ionized paths and streams of electrons that are

Did people in ancient times really believe lightning came from the gods?

The power of lightning has impressed cultures for millennia. People, indeed, felt that it was a divine event and that lightning held mystical powers. The Romans and Greeks viewed them as spears, which were forged by the god Hephaestus (Vulcan, in Roman mythology) and then wielded by Zeus (or Jupiter, in Rome). Norse mythology gives the chief god, Odin, a spear made of lightning that he names Gungnir. And the Hindus also gave their god of war (and weather!), Indra, a thunderbolt weapon. Sometimes, lightning was used as a symbol in these cultures. The Mayans used a triple lightning symbol to represent Huracan, a god of creation who controlled the powers of wind, fire, and storms. There was also a belief, interestingly held by the Mayans, the Romans, and in Hindi culture, that mushrooms would grow in places where lightning struck.

like feelers that the cloud sends out, searching for the best path for the lighting bolt to flow through. Sometimes they may be visible as faintly glowing purplish streaks.

Meanwhile, as the stepped leaders are sending exploring fingers outward, the positively charged ground or other receiving object is sending out feelers upwards. These feelers, called "positive streamers," don't always connect with the stepped leaders, but when they do the circuit is completed and the fireworks are ready to begin. Positive streamers can emerge from any inanimate or living object, including people.

Like railway workers completing a railroad, once the two ends of the path join, the train can drive through. This is the lightning bolt, an explosion of energy that occurs as Nature attempts to equalize the charge between cloud and surface.

Does **lightning** come from **ice**?

Well, in a sense it does. The electric charge that builds up to create lightning is the result of snow and supercooled water droplets rubbing and colliding against each other. This generates a static charge just like when you rub your fuzzy socks against shag carpeting.

Does **lightning** only occur during **thunderstorms**?

Almost always, lightning is associated with thunderstorms. There are a few extraordinary cases, though, where lightning has been observed without storm clouds present. Sometimes, for example, erupting volcanoes can produce smoke plumes that generate the electrical current necessary for lightning; even more rarely, smoke from severe fires can mimic these conditions, as well.

What is a "bolt from the blue"?

This expression relates to the fact that lightning can strike even when there is no rain and the Sun has come out. In other words, blue sky may appear overhead, but there may still be lightning-generating clouds nearby.

Does it have to be **raining** for there to be **lightning**?

There needs to be a storm for there to be lightning, but it does not have to be raining in the same location for lightning to occur. Lightning strokes have been observed as far away as 10 miles (16 kilometers) from where rain was actually falling, and lightning strokes may happen as much as 10 minutes after all rain has stopped.

Has **lightning** ever happened during a **snow storm**?

Yes, it is possible to have lightning and thunder when it is snowing. In fact, winter storms have been known to produce very powerful lightning strokes.

What are the **chances** that I will be **struck by lightning**?

The chance of being struck by lightning at least once in your lifetime is about 1 in 600,000.

Am I **safe** from **lightning** if I **hide indoors**?

Well, you are generally safer to stay indoors, but not 100 percent safe. Most people who are struck by lightning are injured when they are outside (and many injuries and deaths occur when people stand next to trees, so trees are not the best place to find shelter). There have been incidents, however, when people inside their homes have been injured as lightning runs through power lines and pipes and can escape through appliances and plumbing indoors. In one incident that occurred in Chicago Heights, Illinois, on October 24, 1991, a lightning stroke made its way through a cable line, entering a house and causing a bed to catch fire. Such cases are quite rare, however.

How does a **lightning rod** work?

Invented by Benjamin Franklin (1706–1790) around 1750, the lightning rod is designed to provide lightning bolts with a safe path to ground the electricity so that it does not damage a building. Lightning rods have actually become a bit more important to have on a home or other building in recent years, because the metal pipes that used to be installed for indoor plumbing and that could serve as lightning rods are being replaced by nonconductive PVC pipes.

Lightning rods like this one help divert the destructive power of lightning from damaging homes and other structures.

Is **static electricity** similar to lightning?

Yes. When you generate a static charge by, for example, wearing wool socks and rubbing your feet on a shag carpet, and then touching a piece of furniture or a person, the spark that is released is just like a miniature bolt of lightning. Static electricity charges can generate 40,000 volts or more, which as some people know can be damaging to electronics, such as computers.

If I stand next to a **tall object**, will I be **safe from lightning**?

People get this idea from the concept that lightning rods on top of buildings are designed to attract lightning bolts and protect the building from damage. Actually, standing next to a tall object like a telephone pole or tree is no guarantee you won't be hit by the lightning stroke. Lightning will often hit the ground right next to a taller object.

How **hot** is **lightning**?

The temperature of the air around a bolt of lightning is about 54,000°F (30,000°C), which is six times hotter than the surface of the Sun! In cloud-to-ground lightning, energy seeks the shortest route to Earth, which could be through a person's shoulder, down the side of the body, through the leg, and into the ground. As long as the lightning does not pass across the heart or spinal column, the victim can usually survive.

How **bright** is **lightning**?

The light from a lightning bolt is equal to the amount of illumination from about 100 million light bulbs.

How **fast** does **lightning travel**?

The stroke channel of a lightning bolt is very narrow—perhaps as little as half an inch (1.25 centimeters). It is surrounded by a "corona envelope" or a glowing discharge that can be as wide as 10 to 20 feet (3 to 6 meters) in diameter.

How **wide** is a **lightning bolt**?

Because of the intense glow of light that is emitted from lightning bolts, they may appear much wider than they actually are.The speed of lightning can vary from 100

Is it dangerous to touch a person who has just been struck by lightning?

No. A common myth is that people who have been hit by lightning maintain a dangerous electrical charge that can be transmitted to someone else who touches the victim. Actually, no such residual charge will remain and it is safe to touch them and give them medical assistance.

to 1,000 miles (160 to 1,600 kilometers) per second for the downward leader track; the return stroke is 87,000 miles (140,000 kilometers) per second (almost the speed of light).

How many **volts** are in **lightning**?

A bolt of lightning discharges between 10 and 100 million volts of electricity. An average lightning stroke has 30,000 amperes.

How **long** is a **lightning stroke**

The visible length of the streak of lightning depends on the terrain and can vary greatly. In mountainous areas, where clouds are closer to the ground, the flash can be as short as 300 yards (273 meters), whereas in flat terrain, where clouds are higher above the ground, the bolt can measure as long as 4 miles (6.5 kilometers). The typical length is about one mile (1.6 kilometers), but streaks of lightning up to 20 miles (32 kilometers) have been recorded.

How do you **calculate how far away** a **lightning bolt** is?

After you see a flash of lightning, count the number of seconds until you hear the thunder. Divide the number of seconds by five, and the result is the number of miles away that the lightning occurred.

Does lightning give off **X-rays** and **radio waves**?

Almost since the invention of the radio, it has been known that lightning generates radio waves, since this property of the bolts has long interfered with radio broadcasts. Lightning generates radio waves over a wide range of frequencies, especially in the AM broadcast band. More recently, scientists have become fascinated by lightning's ability to create X-rays. The theory that this might occur was first proposed in the 1920s by Nobel Prize-winning physicist C.T.R. Wilson (1869–1959), who theorized that lightning could accelerate electrons with sufficient speed to produce X-rays. For decades scientists thought Wilson was wrong, because they believed the Earth's atmosphere was too thick and that air resistance would slow down electrons too much. However, by the 1990s scientists began to change their minds, and in 2003 experiments with controlled lightning strokes were performed by Martin

Uman at the University of Florida and Joseph Dwyer at the Florida Institute of Technology that demonstrated lightning can, indeed, produce enough energy to overcome any atmospheric drag. Such new research is causing scientists to rethink how lightning works.

How many **times** does **lightning strike** the Earth each **year**?

About 20 million bolts of lightning are generated in the atmosphere every year, and during any particular second about 100 to 125 lightning strokes are occurring on our planet. Thunderstorms are very common in the Earth's atmosphere, with about 1,500 to 2,000 such storms being active at any given time. For astronauts circling the Earth's night side, this makes for an exciting view as numerous white flashes are easily viewed from a space shuttle or from the International Space Station.

Where is **lightning most likely** to occur?

Lightning tends to strike more often over land masses than over oceans, and it is more frequently seen in the tropics, where two-thirds of the electrical storms happen.

What are **keraunophobia** and **brontophobia**?

Keraunophobia is the fear of lightning, and brontophobia—sometimes called tonitrophobia—is the fear of thunder.

How much **energy** does one **bolt of lightning** contain?

A bolt of lighting contains enough energy to light a 100-watt light bulb for three months. To be more technical, each stroke of lightning has about 30,000 amps and one million volts of power, on average. Some "superbolts" can have up to 300,000 amps of power.

Do people frequently **die from lightning strokes**?

Depending on sources, anywhere from 5 to 30 percent of people who are hit by lightning die from their injuries. The danger is not so much from burns but rather from the electrical energy stopping a person's heart. This is why CPR usually is necessary when someone is struck unconscious by lightning. More often, people who have been thus injured will be in severe pain and will be screaming as a result. While they of course will need medical attention, their prognosis for survival is much better than that of the unconscious person.

What is one strange **warning sign** that lightning may be **about to strike**?

As the static charge begins to build as a precursor to a lightning stroke, people's hair may stand on end and you can feel the static charge in the air. It has also been known to happen that a plastic rain coat will begin to rise into the air, or that a cast fishing line will bizarrely remain suspended in air. Any of these signs are a warning to seek immediate shelter.

What did **NASA** and the **U.S. military** discover about **triggered lightning**?

During a couple of space launches, NASA learned that ionized exhaust from rockets can trigger a lightning stroke if rain clouds are nearby. This happened during one of the Apollo missions, though with no dire consequences. More infamously, in 1987 an Air Force rocket launched from the Kennedy Space Center in Florida was hit by lightning. The rocket was destroyed at a cost of $162 million. Scientists have known about triggered lightning for a long time, and they sometimes attach copper lines to small rockets to attract lightning for research purposes. Lightning caused by rocket exhaust, though, is an unintentional side effect.

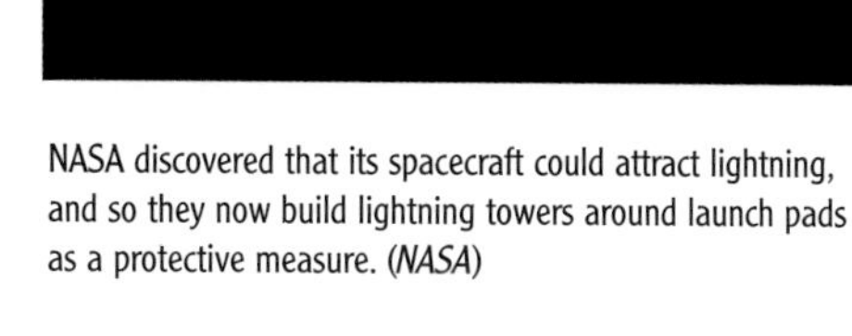

NASA discovered that its spacecraft could attract lightning, and so they now build lightning towers around launch pads as a protective measure. (*NASA*)

During what **part of a thunderstorm's duration** is there the **most risk** of being struck by **lightning**?

Statistics show that more people are hit by lightning strokes toward the end of a thunderstorm. This is not because there is more lightning at that time, but rather because people get too anxious to go outside before the storm is completely over.

How many **people are killed** in the **United States** by **lightning**?

The U.S. total from 1959 to 2003 is 3,696 deaths. About 60 people die each year from lightning; about 300 are injured annually. The highest death rates are in Florida (425 killed between 1959 and 2003), making it the most dangerous state to live in when it comes to lightning injuries. Below is a state-by-state listing using the most recent data from NOAA.

Total Lightning Deaths in the U.S. from 1995 to 2004

State Rank	Number of Deaths	Deaths per Million Residents
1. Florida	85	0.53
2. Texas	34	0.16
3. Colorado	31	0.72
4. Ohio	22	0.19

State Rank	Number of Deaths	Deaths per Million Residents
5. Georgia	19	0.23
6. Alabama	18	0.40
7. Louisiana	17	0.38
7. North Carolina	17	0.21
8. South Carolina	14	0.35
9. Utah	13	0.58
10. Illinois	12	0.10
10. Indiana	12	0.20
10. Pennsylvania	12	0.10
10. Virginia	12	0.17
11. Michigan	11	0.11
12. Oklahoma	10	0.29
12. Tennessee	10	0.18
13. Mississippi	9	0.32
13. Wisconsin	9	0.17
14. Arkansas	8	0.30
15. Arizona	7	0.14
15. Maryland	7	0.13
15. Missouri	7	0.13
15. New Mexico	7	0.38
15. New York	7	0.04
16. Idaho	6	0.46
16. Minnesota	6	0.12
16. Montana	6	0.66
16. New Jersey	6	0.07
16. Wyoming	6	1.21
17. California	5	0.01
17. Iowa	5	0.17
17. Kansas	5	0.19
17. Kentucky	5	0.12
17. West Virginia	5	0.28
18. Nebraska	3	0.18
18. Puerto Rico	3	0.08
18. South Dakota	3	0.40
18. Vermont	3	0.49
19. Connecticut	2	0.06
19. Massachusetts	2	0.03
19. Washington	2	0.03
20. Maine	1	0.08
20. North Dakota	1	0.16
20. Oregon	1	0.03
20. Rhode Island	1	0.10
21. Alaska	0	0

Who are some of the unluckiest people when it comes to being struck by lightning?

U.S. Park Ranger Roy C. Sullivan (1912–1983) survived being struck by lightning seven times between 1942 and 1977. This Virginia ranger was hit once in a car, once in a truck, once while fishing, once while camping, once in his own front yard, once in a ranger station, and once while on top of a lookout tower. Years after his death, he still holds the dubious honor of being the "Human Lightning Rod."

In another peculiar case, a Midwestern family has had numerous members struck by lightning. One woman had been struck by lightning twice—in 1965 and 1995—her grandfather was killed by a 1921 lightning bolt, while her great uncle was also killed in the 1920s; her nephew was temporarily blinded when he was hit; and she also had a cousin who was less seriously injured while holding an umbrella in a storm.

State Rank	Number of Deaths	Deaths per Million Residents
21. Delaware	0	0
21. Washington, D.C.	0	0
21. Hawaii	0	0
21. Nevada	0	0
21. New Hampshire	0	0

Does lightning ever **strike twice** in the same place?

Lightning can and often does strike in the same place twice. Since lightning bolts head for the highest and most conductive point, that point often receives multiple strikes of lightning in the course of a storm—so stay away from something that has already been struck by lightning! Tall buildings (such as the Empire State Building) often receive numerous lightning strikes during a storm.

Can a **lightning stroke** cause you to burst into **flames**?

No, being hit by lightning will not make you combust, but it can leave burn marks or singe clothing.

Can **lightning** travel **through the ground**?

Yes. When lightning reaches the ground, the electrical energy can travel through the ground from some distance away. If you are standing nearby, the energy can then enter through your feet. This is also how multiple people or animals can be injured by a single lightning bolt.

Why does **lightning flicker**?

Once the path is established through which the electricity of the lightning will flow, the same path will be followed for a number of milliseconds before it breaks down. Several strokes of lightning—each lasting millionths of a second—will be interspersed with brief pauses lasting for periods of time just as short. The result is a flickering effect.

How did **lightning cost New York City** a billion dollars?

A lightning stroke took out a major New York City power line in July 1977, resulting in a blackout that lasted for an entire day. The looting and losses to businesses, in addition to the repairs needed, were estimated to cost the city about one billion dollars.

What is a **fulgurite**?

When lightning strikes sandy soil, the soil melts into a glassy stone called a fulgurite. These stones can appear branch- or rootlike, almost as if the bolt of lightning has been fossilized or petrified somehow. The glassy material in the fulgurite is known as lechatelierite, a substance that can also be formed by meteors striking the ground. One of the largest fulgurites ever found is housed at Yale University's Peabody Museum of Natural History and is about 13 feet (4 meters) in length.

What are some **benefits** of **lightning**?

One of the biggest benefits of lightning, believe it or not, is that it causes fires. We usually see fires as a bad thing that destroy plants and property. Lightning starts about 12 percent of all forest fires in the United States, with over 60 percent of these in the Rocky Mountains area and fewer than two percent in the East; most lightning-started fires burn fewer than 10 acres of growth. But botanists and other scientists have long known that fires are beneficial to keeping forests and grasslands healthy. Many plants, indeed, drop seeds that can only germinate after they have been burned, and fires clear away old growth and allow new plants to thrive.

Another benefit to both plants and animals is that lightning strokes convert gaseous nitrogen (N_2) into nitrates (NO_3) by adding energy to the air, which causes nitrogen atoms to bond with oxygen. Nitrates are a vital part of the food chain; plants need them to survive, and animals get them by eating plants or other animals that eat plants. About half the world's naturally occurring nitrates are created by lightning (the other half is generated by bacteria living inside plants such as legumes). Scientists estimate that 200 billion pounds (91 billion kilograms) of nitrates are created *every year* through the action of lightning. In other words, without lightning, plant and animal life on Earth would be severely depleted.

Is it true that being hit by **lightning** can have **health benefits**?

There have been some documented cases where people who were legally blind were hit by lightning and, indeed, found out afterwards that they had regained their

Lightning doesn't always strike the ground; sometimes bolts shoot from cloud to cloud, instead.

vision. There are also a few cases where victims of lightning bolts found out that they did better on intelligence tests afterwards. Indeed, some have even claimed that they have gained psychic abilities.

What are the **different types of lightning**?

Below are descriptions of the various forms that lightning can take.

1. *Normal lightning* (also called streaked or forked lightning) travels from A) cloud to ground, B) cloud to air, C) cloud to cloud, or D) in cloud.
2. *Sheet lightning:* a shapeless flash of lightning that covers a broad area.
3. *Ribbon lightning:* normal lightning blown sideways by the wind in a way that makes it appear like parallel, successive strokes.
4. *Bead or chain lightning:* lightning broken up into evenly spaced segments or beads.
5. *Heat lightning:* lightning seen along the horizon during hot weather that is a reflection of lightning that occurred beyond the horizon in a distant thunderstorm.
6. *Ball lightning:* a rare form of lightning in which a persistent and moving luminous white or colored sphere is seen. Ball lightning can last from a few seconds to several minutes, and it travels at a walking pace. It usually ranges in size from 4 to 8 inches (10 to 20 centimeters), but it has been observed at sizes between 2 inches to 6 feet (5 to 183 centimeters). Because of their unusual

What was the "yellow bubble" lightning seen in 1991?

In Bristol, England, two girls playing Frisbee in 1991 encountered a bizarre "yellow bubble" of energy. It came in contact with them, and both received what felt like an electric shock that threw them to the ground. They lost their breaths for a moment, and upon recovering, ran home and told authorities. No one ever figured out what they saw, but it might have been an unusual form of ball lightning.

behavior, in the past some people have associated them with spirits or other supernatural events. Ball lightning can wander in and out of rooms, usually vanishing harmlessly, but sometimes leaving holes in windows or doors.

What are the **two forms of cloud-to-ground lightning**?

Cloud-to-ground (CG) strokes of lightning come in negative and positive forms. Negative CGs, which make up about 95 percent of all such lightning strokes, occur when the ground becomes positively charged; a positive CG does the opposite, and the ground becomes negatively charged. Positive CGs tend to have more power and longer strike time; thus, they are more likely to cause damage and are blamed for starting more forest fires.

What are some **cloud-to-space forms of lightning** that are not considered to be **true lightning**?

There are four types of electrical phenomena that have been observed that are not really lightning but still involve fascinating atmospheric displays. Called "transient luminous events" or TLE's, they are usually seen during storms. They are sometimes called "cloud-to-space" lightning, though they do not actually originate within clouds. The first scientific paper on these phenomena was published in 1886, but scientists were not very interested in the subject until more recently, as photographic images became increasingly available.

1. Sprites: often reddish lights appearing above thunderstorms for very brief periods of time, sprites look kind of like jellyfish; they have a blob of light on top and numerous tendrils descending downward. Sprites can shoot 55 to 60 miles (about 90 to 95 kilometers) up into the atmosphere, reaching the ionosphere, and extend 100 miles (161 kilometers) across. They are very difficult to see, and for that reason were not reliably recorded until the 1980s.
2. Blue jets: blue lightning that emerges from the tops of thunderstorm clouds at speeds of about 62 miles (100 kilometers) per hour. Meteorologists still do not fully understand what causes blue jets.

3. Elves: a short name for a very long-winded description, elves are emissions of light and very low frequency (VLF) perturbations from electromagnetic pulse (EMP) sources. Appearing as giant rings that expand up to 200 miles (320 kilometers) in diameter, elves exist in the upper atmosphere at elevations of 55 to 60 miles (90 to 95 kilometers). Even more short-lived than sprites, they last about one one-thousandth of a second.
4. Tigers: first observed on January 20, 2003, this newest atmospheric light phenomenon has still not been adequately explained by scientists. Tiger stands for "Transient Ionospheric Glow Emission in Red," and were first observed with the use of an infrared video camera over the Indian Ocean by Ilan Ramon, an Israeli astronaut aboard the space shuttle *Columbia,* which later exploded, killing the crew. The tigers that Ramon observed occurred as bright flashes when there was no thunderstorm activity nearby.

Who first **photographed** a **sprite**?

In 1989, while experimenting with low-light video cameras for use in high-altitude rockets, University of Minnesota scientists Robert Nemzek, John Winckler, and Robert Franz accidentally captured images of sprites. This quickly piqued the interest of NASA scientists at the Marshall Space Flight Center in Huntsville, Alabama, who managed to record more examples of sprites. Later successes of taking pictures or video of sprites came from Walter Lyons, who filmed over 240 sprites on July 7, 1993, and Davis Sentman and Eugene Wescott of the University of Alaska, who used a NASA aircraft to film sprites that same month.

When are you most **likely to see sprites and blue jets**?

The best chance to see one of these brief, strange lights is in the middle of the night, when you are near a strong thunderstorm and far away from the light pollution of a city or town. The storm should be more than 100 miles (161 kilometers) away, but no more than 300 miles (482 kilometers) distant. Estimate the height of the storm clouds, then multiply that by eight to get the approximate altitude where the sprites and blue jets may appear. Sprites may appear as reddish, orange, white, or even greenish flashes; blue jets are even harder to see, but you are more likely to view them if the storm includes hail.

THUNDER AND THUNDERSTORMS

What causes **thunder**?

Thunder is created when lightning rapidly heats a section of air. As the air expands, it compresses, releasing the energy as a sound wave that we hear as a loud boom or clap of thunder.

How do thunderstorms make our planet habitable?

Thunderstorms, of course, usually bring rain or other precipitation with them, which is needed for life on Earth. But other rain storms can do this as well. What makes thunderstorms particularly unique and important is their role in heat convection. Thunderstorms move warm air from lower elevations to upper elevations. The difference in temperatures between ground level and the top of thunderstorm clouds can be as much as 200°F (95°C). Causing this air and temperature circulation helps cool our planet by as much as 20°F (9 to 10°C). Without thunderstorms, therefore, we would already be experiencing global warming on a scale twice as bad as scientists are forecasting because of climate change.

What is a **thunderstorm**?

Thunderstorms are localized atmospheric phenomena that produce heavy rain, thunder and lightning, and sometimes hail. They are formed in cumulonimbus clouds (big and bulbous) that rise many miles into the sky. Most of the southeastern United States has over 40 days of thunderstorm activity each year, and there are about 100,000 thunderstorms across the country annually.

What are the **stages of a thunderstorm**?

Thunderstorms have three stages: cumulus, mature, and dissipation. In the cumulus stage, warm, humid air near the ground is pushed upwards by strong thermals, or by the collision of air masses coming in from several directions at once. As this moist air rises, it cools and the heat that is released enters the surrounding air, causing convection that, in turn, causes more air to rise. A loop that feeds off itself causes a rapid increase in cloud formation, temperature differences between lower and higher altitude, and rainfall. The storm reaches its second stage (the mature stage) when air has reached its cap and is not rising any farther. At this point, cumulonimbus clouds become cumulonimbus incus (thunderheads), with water droplets freezing at the top, thawing into rain as they descend. As the mixture of water, ice, and wind becomes more turbulent, an electrical charge builds up, aligning the ice crystals within the clouds until a bolt of lightning is discharged. This happens repeatedly until the storm weakens in the dissipation stage.

How **loud** is **thunder**?

A clap of thunder can be as loud as 120 decibels (abbreviated dBA), which is as loud as the noise at a rock concert, a chain saw, or a pneumatic drill.

Does thunder cause milk to go sour?

No, this is an old wives' tale. It may have stemmed from the fact that thunderstorms occur in times of heat and humidity, which are conditions that, in themselves, can turn milk sour.

What are the requirements for a storm to be considered a **severe thunderstorm**?

In order for it to be categorized as "severe" a thunderstorm must have winds exceeding 58 miles (93 kilometers) per hour and/or have tornadoes or large hail, or be likely to generate tornadoes or large hail. The National Weather Service issues thunderstorm warnings based on the potential for storms to become severe.

How **tall** can **thunderstorm clouds** be?

Even a fairly average thunderstorm has a height of more than 20,000 feet (6,000 meters), with the tallest storms being reported at over 70,000 feet (21,000 meters).

At **any one time**, how many **thunderstorms are occurring**?

As astronauts orbiting our planet—especially over the night side—can attest, there are thunderstorms happening all over the place, all the time. On average, scientists believe there are about 2,000 active thunderstorms happening at any given time, day or night. This gives us about 16 million thunderstorms every year.

What is the **speed of sound**?

The speed of sound can vary, depending on air pressure and, more importantly, temperature. The conventionally accepted speed, which is useful for estimating the distance to a lightning stroke, is about one mile for every five seconds. More precisely, sound travels at 740 miles per hour (1,191 kilometers per hour) at one atmosphere pressure when the temperature is 32°F (0°C).

Sound travels faster through water and many other media that are denser than air. In general, because the temperature of the air cools as altitude increases, sound is refracted upwards and away from people on the ground. Thus, measurable sound decreases the farther away one is from the source. On the other hand, the stratospheric layer of the atmosphere increases in temperature as altitude increases, thus refracting sound downward.

How far away can **thunder be heard**?

Thunder is the crash and rumble associated with lightning. It is caused by the explosive expansion and contraction of air heated by the stroke of lightning. This results in sound waves that can be heard easily six to seven miles (9.7 to 11.3 kilo-

What is an acoustic shadow?

An acoustic shadow is kind of like a mirage, only involving sound instead of light. Differences in temperatures at varying atmospheric layers causes sound waves to refract or bend; also wind shears and the absorption of sound on soft surfaces can contribute to the effect. The result can be that sounds coming from a particular source might not be heard by someone standing fairly close by, while other people located farther away, but in a direction where sounds are being refracted, *can* hear the sound.

A famous example the consequences of acoustic shadow is often cited from the U.S. Civil War. During the 1862 Battle of Seven Pines in Richmond, Virginia, Confederate General Joseph Johnston told his commanders that, upon hearing gunfire from soldiers being led under Major General D.H. Hill, he would order Brigadier General W.H.C. Whiting to send in an attack on the Union's flank. However, because of acoustic shadow, Johnston was unable to hear the battle noise and the flank attack was not sent in when it was needed. The Confederates subsequently lost the battle.

meters) away. Occasionally such rumbles can be heard as far away as 20 miles (32.2 kilometers). The sound of great claps of thunder is produced when intense heat and the ionizing effect of repeated lightning occurs in a previously heated air path. This creates a shockwave that moves at the speed of sound.

What is **Saint Elmo's fire**?

Saint Elmo's fire has been described as a corona from electric discharge produced on high, grounded metal objects, chimney tops, ship masts, and aircraft wing tips. Since it often occurs during thunderstorms, the electrical source may be lightning. Another description refers to this phenomenon as weak static electricity formed when an electrified cloud touches a high, exposed point. Molecules of gas in the air around this point become ionized and glow. The name originated with sailors who were among the first to witness the display of spearlike or tufted flames on the tops of theirs ships' masts. Saint Elmo (a corruption of Saint Ermo) is the patron saint of sailors, so they named the fire after him.

How do meteorologists **classify thunderstorms**?

There are several classifications of thunderstorms:

Single-cell—the smallest type of storm system, single-cell storms form from a convective loop of warm updrafts and cool downdrafts. They usually form the weakest and briefest rain storms.

Multi-cell—a storm system formed from two or more storm cells.

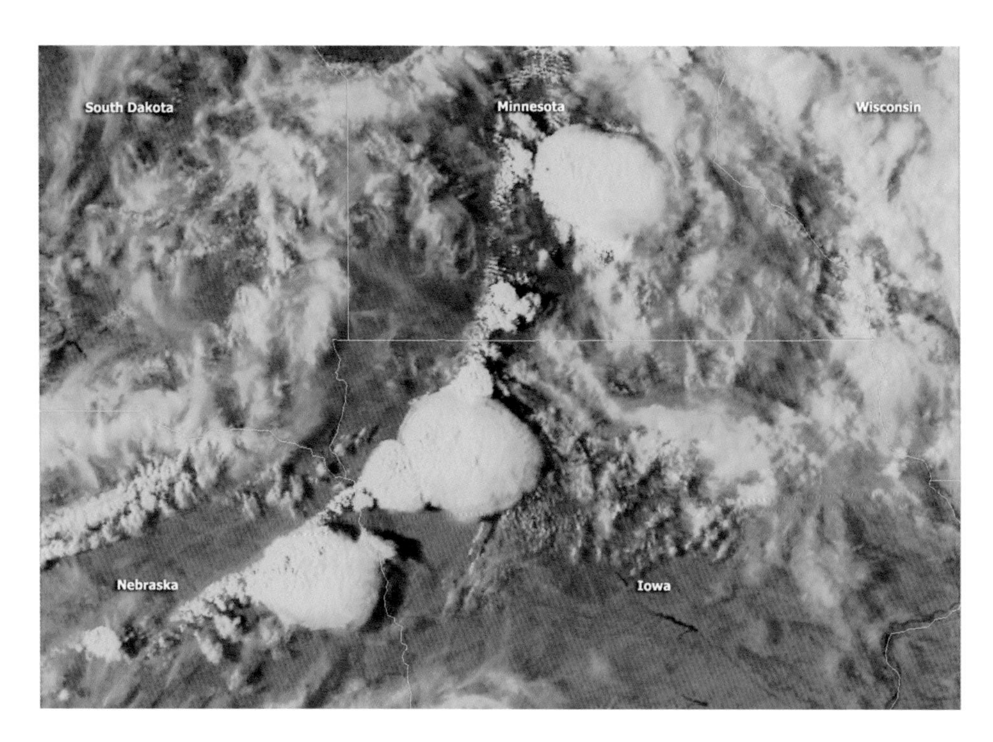

A satellite image from 2001 captures a view of a line of huge supercell thunderstorms moving in a line that extends from Nebraska to Minnesota. (*NOAA*)

Supercell—the largest and most dangerous type of storm system, and one which is often associated with tornadoes. Supercells develop in massive cumulonimbus storm clouds and are characterized by nearly vertical, unsuppressed updrafts and precipitation falling at a nearly horizontal angle. Because the air currents are not suppressed, they tend to continue building in strength for hours.

Squall line—a line of cumulonimbus storm clouds reaching up to 600 miles (965 kilometers) long.

What is an **"indoor thunderstorm"**?

When the humidity is very low inside a building, static electricity can easily build up as people walk across carpeting, dragging their feet, and then touch a piece of furniture, a doorknob, or some other object. When this happens, a discharge of about 40,000 volts can occur in even a small spark jumping from your finger tips. Running a humidifier inside the home during winter can usually keep these "indoor thunderstorms" (a rather exaggerated term) at bay.

Are there **other ways** that **thunderstorms** can be **described**?

Classifying thunderstorms can be a rather subjective practice, but meteorologists do give different types of storms other names based on how they are formed, including:

At this National Severe Storms Laboratory research station, a variety of equipment seen here is used to conduct lightning strike studies. (*NOAA Photo Library, NOAA Central Library; OAR/ERL/National Severe Storms Laboratory*)

Mesoscale convective systems—a large, non-frontal thunderstorm. This classification includes mesoscale convective complexes, which are enormous storm systems that can cover several U.S. states at once and last for half a day or more.

Air mass thunderstorm—a common, short-lived, relatively unorganized, and fairly small storm.

Sea breeze thunderstorm—cold-frontal storms that form as wind blows in toward the coastline.

When do **thunderstorms occur**?

In the United States thunderstorms usually occur in the summertime, especially from May through August. Thunderstorms tend to occur in late spring and summer when large amounts of tropical maritime air move across the United States. Storms usually develop when the surface air is heated the most from the Sun (2 to 4 P.M.). Thunderstorms are relatively rare only in far northern and coastal New England, as well as along the Pacific Coast. Florida, the Gulf states, and the southeastern states tend to have the most storms, averaging 70 to 90 annually. The mountainous southwest averages 50 to 70 storms annually. In the world, thunderstorms are most plentiful in the areas between latitude 35 degrees north and 35 degrees south; in these areas there can be as many as 3,200 storms within a 12-hour nighttime period. As many as 1,800 storms can occur at once throughout the world.

Lightning performs a vital function; it returns to the Earth much of the negative charge the Earth loses by leakage into the atmosphere. The annual death toll in the United States from lightning is greater than the annual death toll from tornadoes or hurricanes—150 Americans die annually from lightning and 250 are injured.

What is a **derecho**?

Pronounced der-RAY-cho, this is large, long-lived thunderstorm characterized by strong wind downbursts.

RAINBOWS AND OTHER COLORFUL PHENOMENA

What is the **solar spectrum**?

The solar spectrum is what results when the Sun's light is broken up by raindrops, ice crystals, or other prisms into its component wavelengths. The shorter wavelengths are in the blue, indigo, and violet spectrum, while the longer wavelengths appear as red, yellow, and orange colors to the human eye.

What is a **rainbow**?

Rainbows are colorful bands of light that are formed when water particles in the air reflect sunlight. As sunlight enters the drops and droplets, the different wavelengths of colors that compose sunlight are refracted at different wavelengths to produce a spectrum of color. To see a rainbow, you must be standing with the Sun behind you and the raindrops in front of you. The Sun needs to be less than 42° above the horizon to obtain the correct angle so that the light waves are properly reflected. The light is refracted as it enters a raindrop, reflects off the inside of the back of the raindrop, and is refracted again as it leaves.

What is the order of the **colors** in a **rainbow**?

The colors in a rainbow are those of a light spectrum: red, orange, yellow, green, blue, indigo, and violet, with the warm colors (red, orange, yellow) on the outside (farthest from the ground) and the cool colors (blue, indigo, violet) closer to the ground.

Who **discovered** how **rainbows form**?

In 1304 a German Dominican monk named Theodoric von Freiburg (1250–1310) showed that light passing through a large glass globe filled with water would create a rainbow because light waves were bent through refraction, reflection, and dispersion. He speculated that the same process occurs in raindrops, creating natural rainbows.

Why do you sometimes see **double rainbows**?

Sometimes rainbows form that have a fainter twin called a "supernumerary rainbow." The twin will be bigger and above the first rainbow, and the colors of the bands will be reversed. This occurs when not all of the light is reflected out of the raindrops at once; some light energy remains, and is reflected within the droplets before emerging as a second rainbow.

Double rainbows sometimes form, with one being slightly fainter than the other, and the colors reversed.

Why is it **impossible** to find a **pot of gold** at the end of a **rainbow**?

Even if there *were* a pot of gold at the end of a rainbow, you could never reach it because the rainbow moves away from you as you approach it. The rainbow will always be positioned opposite the azimuth of the Sun.

Do **rainbows** ever appear by **moonlight**?

Yes, if there are raindrops in the air, and the Moon is full and low on the horizon, the result can be a nighttime rainbow—or "moonbow." Usually, such a rainbow does not have the vibrant hues of its daytime cousin, appearing in muted or whitish shades. A moonbow is a rare and stunning sight.

Is there such a thing as a **monochromatic rainbow**?

There have, indeed, been rare reports of all-red or all-white rainbows.

What **Mie scattering**?

German physicist Gustav Adolf Feodor Wilhelm Mie (1868–1957) discovered how small particles in the atmosphere scatter light waves. Called Mie scattering, this effect is important to meteorologists studying how clouds and haze scatter light (meteorological optics).

How is **Mie scattering** different from **Rayleigh scattering**?

Mie scattering deals with energy waves by particles larger than the wavelengths of light, while Rayleigh scattering (named after English physicist John William Strutt, 3rd Baron Rayleigh [1842–1919]) deals with how light is scattered by particles *smaller* than electromagnetic wavelengths. Mie scattering can explain things like the reddish color of clouds during a sunset (scattering at the level of dust and water particles), while Rayleigh scattering explains why the sky is blue (scattering at the molecular level).

What is a **bishop's ring**?

A bishop's ring is a ring—usually with a reddish outer edge—that is seen around the Sun. It is probably due to dust particles in the air, since it is seen after all significant volcanic eruptions.

A halo, arc, and Sun dog were all photographed simultaneously at a NOAA Antarctic camp. (*photo by Cindy McFee, NOAA Corps*)

What is **irisation**?

Sort of a diffused rainbow effect that can produce mother-of-pearl-like colors, irisation occurs when thin clouds of water vapor (particularly altocumulus clouds) pass below the Sun.

When does the **green flash phenomenon** occur?

On rare occasions, as the last tip of the Sun is setting, the Sun may look bright green for a moment. This green flash occurs because the red rays of light are hidden below the horizon and the blue rays are scattered in the atmosphere. The green rays are seldom seen because of dust and pollution in the lower atmosphere. The phenomenon is best seen when the air is cloudless and when a distant, well-defined horizon exists, such as an ocean view.

What are **Sun dogs** and **Moon dogs**?

When ice crystals are present in the air, the light from the Sun or Moon can be reflected, causing bright spots to appear on either side of these heavenly bodies. Sun dogs—also called "mock suns" or "parhelions"—are sometimes so bright that they look like two companion suns appearing at angles of about 22 degrees on either side of the Sun.

If a **parhelion** is a Sun dog, what is an **anthelion**?

Also caused by hexagonal ice crystals in the air, an anthelion is like a parhelion, except that it is seen at the azimuth opposite the Sun.

What is the Fata Morgana?

This fantastical mirage gets its name from Italy, where this particular type of superior mirage is frequently seen in the Strait of Messina between Sicily and the Italian mainland. Named after the Arthurian legend of Morgan le Fay, who tales tell had a palace beneath the sea, this optical illusion occurs when light diffraction causes distant buildings to appear vertically stretched and, thus, more like castles.

What are **coronas** and **anticoronas**?

A corona is a glowing white ring edged with blue on the inner side and red on the outer, and seen around the Moon, or less often, around the Sun, usually during eclipses. It is caused by light being diffracted by water droplets in a thin cloud, such as an altostratus cloud. The smaller the droplets of water in the air, the larger the corona will appear.

An anticorona—sometimes called a "glory" or "Brocken bow"—is one or more concentric rings that appear when an object, such as an airplane, casts a shadow on a thin layer of clouds. The anticorona will appear on the side opposite of where the light source (Sun or Moon) is.

How is a **corona** different from a **halo**?

As with a corona or anticorona, haloes are created by light passing through clouds of ice crystals. Coronas, when they have some color to them, are reddish on the outside of the ring and bluish on the inside, while haloes are red on the inside and blue on the outside. Coronas tend to be formed when altostratus clouds are present, while haloes occur with cirrostratus clouds in the air. A halo's radius extends at a 22-degree angle from the Moon or Sun, but on rare occasions subtends at 46 degrees; coronas have smaller angular diameters than haloes. Because the cirrostratus clouds causing haloes contain icy crystals that could lead to rain, folklore about the sighting of haloes correctly concluded that haloes may predict oncoming rain.

What is a **Sun pillar**?

Sun pillars are streaks of white or reddish light that extend vertically by as much as 20 degrees from the Sun. Occurring most often at sunrise and sunset, like haloes and coronas Sun pillars are seen because of ice crystals in the atmosphere. In populated areas, when icy fog occurs Sun pillars can sometimes be seen around street lights.

What is a **crepuscular ray**?

A less technical name for a crepuscular ray would be a sunbeam, such as can be seen when streams of sunlight emerge from behind a cloud during the twilight hours

Crepuscular rays stream down like a like light beams during an Antarctic sunset. (*photo by Dave Mobley, Jet Propulsion Laboratory, courtesy NOAA*)

("crepuscule" means "twilight"). In the Bible, they may have inspired the story of Jacob's ladder (Genesis 28:11-19), and so they are often named after that story. Another descriptive phrase for crepuscular rays is "the Sun drawing water" because people once believed that the beams of light were actually formed by water being sucked up into the Sun. Because these rays are seen as clouds are breaking up, folklore rightfully interprets them as a sign that good weather will likely be ahead.

What is a **mirage**?

As light passes from warm to cool air, or vice versa, light is refracted, causing images to appear in places distant from their actual location. Mirages are frequently seen on hot days and on surfaces such as road pavement, sand, or concrete, where the surfaces are very hot. In such cases, light rays from the distant horizon appear wavy and close up, making it appear as if water is nearby.

What is a **superior mirage**?

A superior mirage—also called a "superior image"—refers to the illusion that occurs when light from objects is refracted in a way that makes them appear as if they are floating in air above the actual object. This can occur when warm air blows over cooler surfaces (the opposite of a common mirage).

What **military battle** was **cancelled** because of a **mirage**?

During World War I, a battle between British and Ottoman forces in April 1916 was cancelled because a mirage made it impossible for the combatants to see one another.

A vivid Aurora Borealis is seen over Fairbanks, Alaska.

What is an **aurora**?

An aurora is a bright, colorful display of light in the night sky. Aurorae are produced when charged particles from the Sun (usually solar wind particles, but sometimes coronal mass ejections as well) enter Earth's atmosphere. The particles are guided to the north and south magnetic poles by Earth's magnetic field. Along the way, these particles ionize some of the gas molecules they encounter by drawing away electrons from those molecules. When the ionized gas and their electrons recombine, they glow in distinctive colors; and the glowing gas undulates across the sky.

What is the difference between the **Aurora Borealis** and the **Aurora Australis**?

The Aurora Borealis is the name for aurorae that appear in the Northern Hemisphere, while the Aurora Australis appears in the Southern Hemisphere.

Where are **aurorae** seen?

Aurorae are most prominent at high altitudes near the North and South Poles. They can also be seen sometimes at lower latitudes on clear nights, far from city lights; every once in a while—perhaps once every year or so—aurorae can be seen as far south as the lower 48 states of the United States. Displays of aurorae can be amazingly beautiful, varying in color from whitish-green to deep red and taking on forms like streamers, arcs, curtains, and shells.

How **often** does an **aurora** appear?

Because it depends on solar winds and sunspot activity, the frequency of an aurora cannot be predicted. Auroras usually appear two days after a solar flare (a violent eruption of particles on the Sun's surface) and reach their peak two years into the 11-year sunspot cycle.

Is the **best time** to see an **aurora** during the **winter**?

Seasons don't have anything to do with whether or not aurorae will occur over the polar regions. Rather, they are related to solar storms. However, it may be more likely that you will see aurorae during the winter because the nights are longer.

Can you see the **aurorae** during the **day**?

No, human eyes can't detect aurorae during the daytime. However, the aurorae *are* there, and satellites that can detect X-rays, such as the POES, are able to monitor aurorae activity.

What is **airglow**?

If you could remove all the illumination coming from city lights, the stars and Moon, and from aurorae, the night sky would still have a faint, greenish glow to it. This is caused by oxygen atoms about 60 miles (100 kilometers) high being excited by radiation from space. Other colors coming from different elements are also possible, but the green from oxygen predominates. This phenomenon is sometimes called the "permanent aurora."

GEOGRAPHY, OCEANOGRAPHY, AND WEATHER

Why is it important to understand **geography** in relation to the **weather**?

There are a couple of reasons why geography is an important science for meteorologists to understand. One is that geographical features, such as mountains and coastlines, have an important influence on the weather. The other is that meteorologists interpret data from satellites, radar, and computer projections on various types of maps, and it is important for them to be able to visualize this data in three-dimensional space because the planet is not flat.

PLATE TECTONICS

Who **initially proposed** the idea of **moving continents**?

The idea of the continents moving around our planet was mentioned as early as 1587 by the Flemish map maker (with German origins) Abraham Ortelius (1527–1598) in his work *Thesaurus geographicus*. In 1620, Francis Bacon (1561–1626) also mentioned the idea, noting the fit of the coastlines on both sides of the Atlantic Ocean. By the 1880s, many other scientists were mentioning the connection. For example, in 1885, Australian geologist Edward Seuss (1831–1914) proposed that the southern continents had once been a huge landmass that he called Gondwanaland.

But it was German scientist Alfred Wegener (1880–1930) who first formally published the idea of continental displacement (or drift) in his 1915 book, *The Origins of Continents and Oceans*. He believed the continents were once joined together into one supercontinent, a place he named Pangaea (also spelled Pangea, meaning "all land") that was surrounded by a superocean called Panthalassa. He also suggested that the massive continent divided about 200 million years ago, with Laurasia moving to the north and Gondwana (or Gondwanaland) to the south. Wegener based his ideas of continental motion on numerous observations: The continental distribution of fossil ferns called *Glossopteris* (from studies by Seuss); the discovery of coal in Antarctica by Sir Ernest Henry Shackleton (1874–1922); similar glacial erosion seen in the tropical areas of India, South Africa, and Australia; the apparent fit of the South America and west African continental shorelines; and, although it may only be legend, by watching ice floes drifting on the sea.

Although Wegener is now considered "the man who started a revolution" in geology, his ideas were hotly debated by scientists of his time. Not only was he a meteorologist in a community of geologists, but he could offer no logical mechanism for the movement of the landmasses. It wasn't until the 1960s, long after his tragic death in Greenland (he died at the age of 50 while on a rescue mission), that Wegener was vindicated. By then, scientific measurements, observations, and technology had advanced enough to prove that, indeed, the continents are moving around the planet on giant lithospheric plates. Wegener's theory of continental displacement was replaced by the new field of plate tectonics, which is the basis for modern geology.

So, what is the modern theory of **plate tectonics**?

The Earth's crust and lithosphere are broken into over a dozen thin, rigid shells, or plates, that move around the planet over the plastic aesthenosphere in the upper mantle. The interaction between these plates is called tectonics, from the Greek *tekon* for "builder"; plate tectonics describes the deformation of the Earth's surface as these plates collide, pass by, go over, or go under each other. In other words, plate tectonics describes how these plates move, but not why.

Overall, plate tectonics combines Wegener's theory of continental displacement (or drift) and Hess's discovery of seafloor spreading (see below). The theory has truly revolutionized the study of the Earth's crust and deep interior. It allows scientists to study and understand the formation of such features as mountains, volcanoes, ocean basins, mid-ocean ridges, and deep-sea trenches, and to understand earthquakes and volcano formation. It also gives clues as to how the continents and oceans looked in the geologic past, and even how the climate and life forms evolved.

What **physical evidence** shows that the **continents move**?

Scientists have gathered plenty of evidence that shows the continents move over time. For example, the shape of the continents and their fit was determined by Sir Edward Bullard in 1965. He did not site the usual continental shapes we see, but he measured the "real" edge of the continents: the continental slope, an area that shows a much better fit at the 6,560 foot (2,000 meter) depth contour than at the shorelines of continents.

Other scientists matched the continental geology on either side of an ocean. For example, the mountain belts of the Appalachians and the Caledonides are relatively similar geologically, as are the sedimentary basins of South Africa and Argentina. Another way to prove that continents move over time includes paleontology, in which similarities or differences of fossils on certain continents indicate

> **What is the connection between earthquakes and plate tectonics?**
>
> Only the lithosphere has the strength and brittle behavior to fracture in an earthquake. And as lithospheric plate boundaries push, pull apart, or grind against each other, earthquakes occur. In 1969, scientists published the locations of all earthquakes that occurred from 1961 to 1967. They discovered most earthquakes (and volcanoes, too, they later learned) occurred in narrow belts around the world. Thus, it is now known that areas with frequent earthquakes and volcanoes help define the plate boundaries.

a match. For example, there are similar Mesozoic Era reptiles in North America and Europe, a time when scientists believe those two continents were joined together; similar Carboniferous and Permian flora and fauna are found in South America, Africa, Antarctica, Australia, and India. In contrast—no doubt after the continents were well separated—there is a wide diversity of organisms in the Cenozoic Era.

Who contributed to **early work in plate tectonics**?

There were several key scientists who contributed to the study of plate tectonics as it became more favored in the late 1960s. One of the most popular scientists to discover evidence for plate tectonics was J. Tuzo Wilson (1908–1993). By 1965, he described the origin of the San Andreas fault, the large crack in the Earth's surface near San Francisco, California, as a transform fault (or strike-slip)—one of the major plate boundaries. In 1968, Xavier LePichon (1937–) participated in the definition of the overall "plate tectonics" model and published the first model quantitatively describing the motion of six main plates at the Earth's surface; in 1973, he wrote the first textbook on the subject.

Other geologists have made major contributions to the development of the plate tectonics theory: William Jason Morgan published a landmark paper in 1968 explaining the many tectonic plates and their movements; he also recognized the importance of mid-plate volcanic hot spots that create island chains such as the Hawaiian Islands. Walter Pitman III was instrumental in interpreting the pattern of marine magnetic anomalies detected around mid-ocean ridges, an indicator of active seafloor spreading and evidence of plate tectonics. And Lynn R. Sykes used seismology to refine plate tectonics, and he noted the connection between transform faults at the mid-ocean ridges and plate motion. He also coauthored *Seismology and the New Global Tectonics* in 1968, which relates how existing seismic data could be explained in terms of plate tectonics.

What did the **continents** look like **in the past**?

Because of the movement of the lithospheric plates, the continents' positions have changed over time. For example, some scientists believe that about 700 million years ago a huge continent called Rodinia formed around the equator; about 500 million years ago, the continent broke apart, forming Laurasia (today's North America and Eurasia) and Gondwana (or Gondwanaland; today's South America, Africa, Antarctica, Australia, and India). Then, about 250 million years ago, the continents were once again together in one massive supercontinent called Pangea (or Pangaea, translated as "all land"). Eventually, the huge continent began to break up, forming Laurasia and Gondwana again.

What is **continentality**?

Areas of a continent that are distant from an ocean (such as the central United States) experience greater extremes in temperature than do places that are closer to an ocean. These inland areas experience continentality. It might be very hot during the summer, but it can also get very cold in winter. Areas close to oceans experience moderating effects from the ocean that reduce the range in temperatures.

What is **seafloor spreading**?

Seafloor spreading is one of the processes that helps move the lithospheric plates around the world. The process is slow but continuous: Like a hot, bubbling stew on the stove, the even hotter asthenospheric mantle rises to the surface and spreads laterally, transporting oceans and continents as if they were on a slow conveyor belt. This area is usually called a mid-ocean ridge, such as the Mid-Atlantic Ridge system in the Atlantic Ocean.

The newly created lithosphere eventually cools as it gets farther from the spreading center. (This is why the oceanic lithosphere is youngest at the mid-ocean ridges and gets progressively older farther away.) As it cools, it becomes more dense. Because of this, it rides lower in the underlying asthenosphere, which is why the oceans are deepest away from the spreading centers and more shallow at the mid-ocean ridges. After thousands to millions of years, the cooled area reaches another plate boundary, either subducting, colliding, or rubbing past another plate. If part of the plate subducts, it will eventually be heated and recycled back into the mantle, rising again in millions of years at another or the same spreading center.

How was **seafloor spreading discovered**?

In the 1950s, scientists realized that as igneous rocks cool and solidify (crystallize), magnetic minerals align with the Earth's magnetic field like tiny compass needles, essentially locking the magnetic field into the rock. In other words, rocks with magnetic minerals act like fossils of the magnetic field, allowing scientists to "read" the rock and determine the magnetic field from the geologic past. This is called paleomagnetism.

The idea was proposed by Harry Hess (1906–1969), a Princeton University geologist and U.S. Naval Reserve rear admiral, and independently by Robert Deitz, a scientist with the U.S. Coast and Geodetic Survey, both of whom published similar theories that became known as seafloor spreading. In 1962, Hess proposed the idea of seafloor spreading, but had no proof. As Hess formulated his hypothesis, Dietz independently proposed a similar model, which differed by noting the sliding surface was at the base of the lithosphere, not at the base of the crust.

Support for Hess's and Dietz's theories came only one year later: British geologists Frederick Vine and Drummond Matthews discovered the periodic magnetic reversals in the Earth's crust. Taking data from around mid-ocean ridges (seafloor spreading areas), Vine noted the magnetic fields of magnetic minerals showed reversed polarity. (The Earth's magnetic field has reversed its polarity around 170 times in the last 80 million years.) From the spreading center outward, there was a pattern of alternating magnetic polarity on the ocean floor—swaths of opposing polarity on each side of the ridge. As the spreading center continues to grow, new swaths develop, pushing away material on either side of the ridge. Thus, these strips of magnetism were used as evidence of lithospheric plate movement and of seafloor spreading.

How **fast** does the **seafloor spread**?

Today, the rates of seafloor spreading vary from about 1 inch (2.54 centimeters) per year in the mid-Atlantic ridge area to about 6 inches (15 centimeters) in the mid-Pacific Ocean. Scientists believe seafloor spreading rates have varied over time. For example, during the Cretaceous Period (between 146 to 65 million years ago) seafloor spreading was extremely rapid. Some researchers believe this quick movement of the lithospheric plates may have also contributed to the demise of the dinosaurs: As the continents changed places over time, so did the climate. In addition, more plate movements might have meant more volcanic activity, releasing dust, ash, and gases into the upper atmosphere and contributing to more climate variation. This change in climate and vegetation may have cause several species of dinosaurs to die out or become diseased, contributing to the dinosaurs' extinction.

RAIN, ICE, AND GEOGRAPHY

Why is it **wetter** on **one side** of a **mountain range**?

It's much more wet on one side of a mountain than the other because of a process known as orographic precipitation. Orographic precipitation causes air to rise up the side of a mountain range and cool, creating precipitation and storms. The storms deposit a great deal of precipitation on that side of the mountain and create a rain shadow effect on the opposite side of the range. The Sierra Nevada mountains are an excellent example of orographic precipitation because the mountains of the western Sierras receive considerable rainfall (far more than California's Central Valley), while the eastern Sierras of Nevada are quite dry.

A satellite image of Greenland shows a large island almost entirely covered by a thick sheet of ice. (*NASA*)

What is a **rain shadow**?

When the moisture in the air is squeezed out by orographic precipitation, there's not much left for the other side of the mountains. The dry side of the mountains experiences a rain shadow effect.

Why is **Greenland** called **green** when it is mostly **covered in ice**?

When Scandinavian explorers first discovered the large island north of Iceland in the late-tenth century, they wished to attract more settlers to the land, and so they named it Greenland, according to some sources. Another explanation is that the island was actually named "Gruntland," the word "grunt" meaning "shallow bay." The name was later mistranslated on maps, becoming Greenland. While much of the island is inhospitably covered by a huge glacier, the southern coastline does actually have vegetation and has served as good fishing ground. The Little Ice Age of the fifteenth century decimated the Viking settlements, however.

Which continents have the **highest and lowest average elevations** on the planet?

Antarctica has the highest average elevation (7,546 feet [2,300 meters] above sea level), much of which is due to the permanent glacial ice. Australia has the lowest average (984 feet [300 meters] above sea level).

How much **ice** is there covering **Antarctica and Greenland**?

Most of the world's fresh water is stored in the form of ice covering the continent of Antarctica: an amazing 70 percent! Furthermore, 90 percent of all the Earth's ice is also found in the southernmost continent. The ice covering Antarctica is as much as two miles (3.2 kilometers) thick in places. Scientists have found that the ice is getting deeper in the central part of the continent, while on the outer edges it seems to be melting. Although scientists do not understand why yet, the ice on the western side of the continent seems to be getting thicker overall, while in the east it is getting thinner. Overall, the total amount of ice appears to be at stable levels, in contrast to ice in the Northern Hemisphere. It is difficult to say how much ice is covering Greenland, because there it is melting so quickly. In the late 1990s, the large island had an ice cap that contained about 720,000 cubic miles (three million cubic kilometers) of ice. At that time, the ice was melting at a rate of about 21.6 cubic miles (90 cubic kilometers) per year. As of 2005, that rate has increased to 36 cubic miles (150 cubic kilometers).

Where does it **snow on the equator**?

Mountainous regions on or near the equator regularly get snow. For example, it snows in the Andes Mountains in Ecuador, and in Africa snow falls on Mt. Kenya and Mt. Kilimanjaro.

It is possible to find snow near the equator, such as atop Mt. Kilimanjaro in Tanzania.

VOLCANOES

What are the **two major characteristics** of volcanic **eruptions**?

A volcano can experience violent eruptions or less explosive, more effusive eruptions that produce wide-ranging lava flows. In general, the eruptive characteristics of a volcano are based on the silica and water content of the magma.

What **major gases** do **volcanoes emit**?

Volcanic gases contained within the magma (molten rock) are released as they reach the Earth's surface, escaping at the major volcanic opening or from fissures and vents along the side of the volcano. The most prevalent gases are carbon dioxide (CO_2) and hydrogen sulfide (H_2S). Carbon dioxide is a dangerous gas; it is invisible and odorless, and can kill within minutes.

One example in which volcanic gases proved dangerous involved the Dieng Volcano Complex (or Dieng Plateau) in Java, Indonesia. It consists of two volcanoes and over 20 craters and cones, and is noted for its poisonous gas emissions at some craters. In 1979, at least 149 people were killed by poisonous gases as they fled eruptions at two of the craters: the Sinila and Sigludung.

Did **volcanoes** play a role in **creating Earth's atmosphere**?

Scientists now believe that much of our planet's atmosphere was generated by carbon dioxide, water vapor, nitrogen, argon, and methane spewing out of volcanoes. When life began to form as primitive plant cells, the carbon dioxide issued from volcanoes was absorbed by these plants and then released as oxygen. At first, the oxygen reacted with iron and other metals in the Earth's crust, creating iron oxides that form the commonly seen reddish earth in the ground. Eventually, though, there was enough oxygen that it became part of the atmosphere, and breathable air was created.

Can **volcanic eruptions affect** the global **climate**?

Most volcanic eruptions do not affect the global climate, although larger ones can cause disruptions—albeit for a relatively short period of time. Large eruptions tend

An ash plume erupts from the Cleveland Volcano in Alaska's Aleutian Islands on May 23, 2006. International Space Station flight engineer Jeff Williams took this photo after warning the Alaska Volcano Observatory of the activity. (*NASA*)

to eject gases and dust high into the stratosphere. From there, prevailing winds carry the particles around the world—sometimes with interesting results.

For example, in 1815, Mt. Tambora on the island of Sumbawa (near Java, Indonesia) erupted, putting out a record amount of ash that briefly changed the world's climate. Huge amounts of volcanic dust rose high into the atmosphere, reaching around the globe. That year (and for some of the following year), volcanic particles screened out some sunlight, causing the global temperatures to fall. In Europe and other parts of the Northern Hemisphere, winter never seemed to end, with frosts occurring throughout the summer. Hence, 1816 is known as "the year with no summer."

Scientists used to believe that cooling of the atmosphere could result from volcanic eruptions because of the amount of ash that was thrown up into the air. Now they know, however, that most of these fine particulates return to the Earth within about six months. What actually has a greater effect is the sulfur dioxide (SO_2) that volcanoes produce. Sulfur dioxide reacts with water vapor, and the result is a long-lasting haze that blocks out a considerable amount of the Sun's radiation.

Who first theorized that there was a **link between volcanoes and climate**?

As well as being one of the Founding Fathers of the United States, an inventor, and a diplomat, Benjamin Franklin (1706–1790) is often credited as the first person to notice that volcanic activity might be affecting weather. Observing that, after the eruption of Iceland's Laki volcano in 1783, there seemed to be a period of cooler weather lasting into 1784, Franklin believed that the more common incidence of fogs in Europe was a consequence of the eruption.

How **many active volcanoes** are there?

Currently, there are somewhere between 850 and 1,500 known active volcanoes on our planet; there are 63 active volcanoes in the United States, mostly in Alaska, Hawaii, and in the Pacific Northwest. At any given time, about 10 or 12 volcanoes are erupting planet wide.

What is a **fumarole**?

Volcanic gases escape from fumaroles, or vents, around volcanically active areas. They can occur along tiny cracks or long fissures in a volcano, in groups called clusters or fields, and on the surfaces of lava and pyroclastic flows. Fumaroles have been known to last for centuries. They can also disappear in a few weeks or months if their source cools quickly. For example, Yellowstone National Park and the Kilauea volcanoes have many fumaroles and associated deposits; some have been there for years, while others have just recently appeared.

What is **tephra**?

Tephra is the name given to all the material that erupts from a volcano, excluding lava. Tephra comes in all shapes and sizes, and is also referred to as pyroclastic material ("fire particles"). A pyroclast is material that is ejected during the explosive eruption of a volcano in the form of fragments; pyroclastic material that is hot enough to fuse together before it falls to the ground is called welded or volcanic tuff. Geologists classify tephra according to size. The following lists the most common types of tephra:

Ash—Ash is material smaller than approximately a tenth of an inch (2 millimeters) that is emitted from an erupting volcano; it can also contain lapilli (also called cinders or "little stones"), which is between 1 and 25 inches (2 and 64 centimeters). In a large eruption, ash can accumulate to a great thickness and spread out for thousands of miles (usually in the direction of the prevailing winds).

Block—Blocks are solid rock emitted from an erupting volcano. They can be anywhere from the size of a baseball to the size of a boulder as large as a house.

Bombs—Bombs are volcanic rocks that are still molten inside; they are shaped by their passage through the air. (They form the brilliant arcs seen in time-lapse photography of volcanic eruptions.) Typically ranging from baseball to basketball size, they can be as large as a house. Bombs (and blocks) can be ejected from a volcano with initial velocities greater than 1,000 miles (1,609 kilometers) per hour, and can travel more than 3 miles (5 kilometers), with some exploding and gushing molten rock when they eventually strike the ground. There are also certain types of bombs, including spindle bombs (very fluid magma chunks that are

The scenic geysers and hot springs of Yellowstone are the result of volcanic activity just under the surface of the famous national park.

streamlined as they fly through the air) and bread crust, which is formed from viscous magma, creating rounded blobs that often have fractured surfaces.

What is a **phreatic eruption**?

Phreatic eruptions are steam-driven explosions that occur when groundwater or surface water is heated by magma, lava, hot rocks, or new volcanic deposits. The intense heat of such material can cause water to boil and turn to steam, generating an explosion of steam, water, ash, blocks, and bombs.

How much **volcanic activity** occurs **underwater**?

There is a huge amount of volcanic activity taking place underwater—we just can't see it. Some geologists estimate that approximately 80 percent of all Earth's volcanic activity occurs on the ocean floor.

What are **black smokers**?

Black smokers are actually deep-ocean hydrothermal (hot water) vents, named after the dark, soot-like material ejected from "chimney" formations on the ocean floor. The material is actually superheated water (around 662°F [350°C]) with very high concentrations of dissolved minerals—mostly sulfur-bearing minerals or sulfides from lava on a mid-ocean ridge volcano. As the hot water meets the cold ocean waters, the minerals precipitate out, settling out around the surrounding rock. Over time, the hollowed-out chimneys grow taller as more minerals precipitate out.

Black smokers tend to occur in volcanic vent fields that are typically tens of yards across, with fields ranging from pool-table size (43 square feet [4 square meters]) to tennis court size (8,288 square feet [770 square meters]). For example, vent fields are found on the Juan de Fuca Ridge, a mid-ocean ridge in the Pacific Ocean. Many vents have been discovered since the first site was found in 1977 near the Galapagos Islands (in the small research submersible *Alvin*), and there are probably many more. But scientists have only explored a small portion of the Earth's mid-ocean ridges.

What caused a temporary but dramatic **change in climate in 1816** called the **Year without a Summer**?

On April 10, 1815, Mount Tambora erupted on the Indonesian island of Sumbawa after what some scientists believe was a period of inactivity of about 5,000 years. The

Is Yellowstone actually a supervolcano?

Yellowstone National Park is renowned for its scenic geysers and hot springs covering some 2,300 square miles (3,800 square kilometers). The energy for all this hot water actually comes from volcanic energy below a plateau that is actually a giant caldera. Geologists estimate that this massive volcano last erupted about 640,000 years ago with an energy equal to 8,000 Mount St. Helens eruptions. One can only imagine the destruction such an explosion would have wreaked. A nuclear winter would have ensued, acid rain would have fallen from the sky, and some scientists believe that the human race was driven almost to extinction at the time.

Since the volcano is still active, it could happen again at almost any time. Some parts of Yellowstone have risen about 29 inches (74 centimeters) since measurements were first taken in 1923, which indicates a buildup of magma beneath the crust. It is only a matter of time before this energy is released. And Yellowstone is not the only such supervolcano. The last such eruption came from Toba, Sumatra, about 70,000 to 75,000 years ago, and other, as-yet-undiscovered supervolcanoes might exist.

massive blast shot an estimated 24 cubic miles (100 cubic kilometers) of rock and 220 million tons (199.54 billion kilograms) of sulfur dioxide almost 29 miles (40 kilometers) into the atmosphere. Everything within 400 miles (645 kilometers) of the blast was plunged into near total darkness. Tsunamis reaching 16 feet (five meters) in height crashed onto surrounding shorelines. After the initial eruption, the volcano continued to erupt through mid July.

To make matters worse, two previous volcanic eruptions also contributed to dust and debris being thrown into the Earth's atmosphere. The volcano Soufrière St. Vincent erupted in the Caribbean in 1812, and two years later Mayon Volcano exploded in the Philippines. In addition to all this volcanic activity, there was an increase in sunspot activity, including one particularly large sunspot that was so big it could be seen with the naked eye.

All of these events conspired to lower temperatures dramatically all over the planet. Canada, Europe, and the United States were all very hard hit, particularly in the East and Midwest, where crop failures were extensive in 1816. During that summer, snow fell a few times in New England in June. Vermont saw ice on its lakes in June thick enough to skate on, while snow was as deep as 18 to 20 inches (45 to 50 centimeters) deep. After this bizarrely cold summer, the winter of 1816 to 1817 was rather mild, but that did not matter since the crops were all ruined.

What happened when **Mt. St. Helens erupted**?

On the morning of May 18, 1980, a magnitude 5.1 earthquake struck beneath the Mt. St. Helens volcano in Washington state. Later that day, the volcano exploded,

Smoke emerges from Mt. St. Helens before its 1980 eruption.

initiating a massive avalanche that tore away the northern slope of the mountain and created the largest landslide in recorded history. The conical volcano went from about 9,678 feet (2,950 meters) to 8,366 feet (2,550 meters) in height, releasing a giant plume of ash and gas high into the atmosphere. A lethal pyroclastic flow of hot steam, gas, and rock debris raced down the slope of the mountain, traveling as fast as 684 miles (1,100 kilometers) per hour.

A fine ash dust was propelled into the upper atmosphere (stratosphere) at heights reaching up to 15 miles (22 kilometers), spreading to the east by the prevailing westerly winds and eventually reaching all over the world. In a short time, the ash blanketed central Washington, the prevailing winds carrying an estimated 540 million tons of ash across 22,007 square miles (57,000 square kilometers) of the western United States. People in towns nearby (and some as far away as western Montana) were affected by the rain of ash. Car radiators were clogged, upper respiratory problems worsened, air and ground travel were disrupted, and a thick coating of ash particles covered everything outside. Despite such impressive statistics, the Mt. St. Helens eruption did not disrupt climate on a worldwide scale. There are two reasons for this: 1) the volcano did not release that much sulfur dioxide compared to some other major eruptions, and 2) the eruption blew out the side of the mountain, thus shooting debris at an angle and not as high into the atmosphere as might have otherwise happened.

What **two volcanic eruptions** had the biggest **impacts on the climate** in the **twentieth century**?

The eruption of El Chichón in southern Mexico, which lasted from March 29 through April 4, 1982, and the June 15, 1991, eruption of Mt. Pinatubo in the Philippines caused significant disruptions to the planet's climate. El Chichón shot about 7.75 million tons (over 7 billion kilograms) of sulfur dioxide into the atmosphere, as well as some 24.25 million tons (22 billion kilograms) of other dust and particles. Coincidentally, there was a strong El Niño building at the same time. While the El Niño effect worked to warm ocean waters, the El Chichón eruption was cooling the atmosphere, and the result was that the two effectively cancelled each other out. That summer, when temperatures should have increased because of El Niño, the average temperatures were actually fairly normal. During the winter of 1982 to 1983, though, temperatures in Europe, Siberia, and North America were higher than normal, and temperatures in the Middle East, China, Greenland, and Alaska were cooler. This was because the gases from El Chichón had caused an arctic oscillation in the stratosphere, changing air current patterns.

When Mount Pinatubo erupted, it sent 20 million tons of sulfur dioxide into the sky, and estimates are that this resulted in an average worldwide temperature drop of 1.7°F (0.8°C) in 1992. The effects continued through 1993, as the haze produced by the extra sulfur dioxide in the atmosphere reflected the Sun's rays.

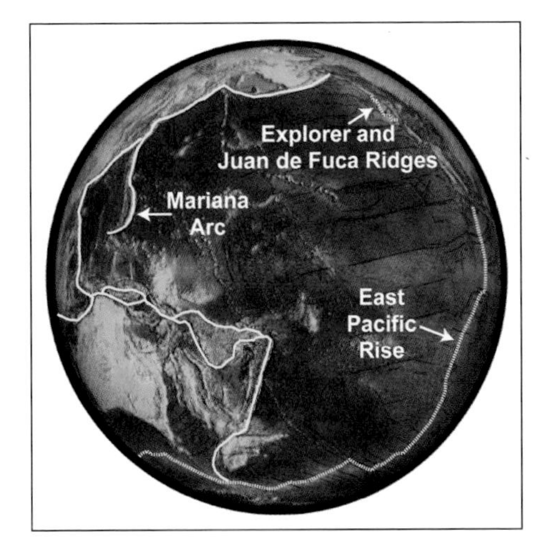

This map of the Ring of Fire indicates the Mid Ocean Ridge Systems with a dotted line, and the Island Arc/Trench Systems with a solid yellow line. (*Pacific Ring of Fire 2004 Expedition. NOAA Office of Ocean Exploration; Dr. Bob Embley, NOAA PMEL, Chief Scientist*)

What is the **Ring of Fire**?

There is a circular region that surrounds the Pacific Ocean where volcanic activity is particularly high. This is known as the "Ring of Fire" and includes coastal areas in Japan, Russia, Alaska, Canada, Oregon, Washington state, California, Mexico, Southeast Asia, and many South Pacific islands. The ring stretches some 40,000 miles (64,000 kilometers) and includes three-fourths of the planet's volcanoes. Among those are Mt. St. Helens and recently active volcanoes in Alaska, such as Mount Spur, which erupted in 1992, and Mt. Redoubt, which erupted March 22, 2009, near Anchorage.

OCEANOGRAPHY AND WEATHER

What is **oceanography**?

Oceanography is the study of the world's oceans, including the waters and everything in them: animals, plants, and minerals. Oceanographers study the physics, chemistry, biology, and geology of the seas. Oceanography is important to understand in relation to meteorology for many reasons. For example, the oceans have a lot to do with heat absorption, distribution, and reflection, as well as with the water cycle and with levels of carbon dioxide in the atmosphere, among other influences.

How much **water** is there on **Earth**?

Including all the world's oceans, lakes, rivers, as well as all the water contained in the Earth's soils, in the atmosphere, and in icebergs and other frozen forms, scientists estimate that there is 3.7×10^{14} gallons (1.4×10^{15} liters) of water on the planet.

How much **water** is in the **world's oceans**?

Earth is about 70 percent covered by oceans and seas, and about 97 percent of the world's total water is contained in the oceans. Two percent of this water is in the form of ice.

What is a **hydrometer**?

A hydrometer measures the specific gravity of a liquid. It is used to determine the density of a fluid compared to the density of pure water at 60°F (15.5°C). This can be handy when seeking a reading for the salinity of water, such as when taking samples of sea water.

How much **water evaporates** from the world's **oceans**?

Incredibly, over 1.32×10^{17} gallons (500,000 cubic kilometers) of water evaporate from the oceans each year. Fortunately, that water is replenished by 1.19×10^{17} gallons (450,000 cubic kilometers) of rain and snowfall, as well as waters draining into the oceans and seas from rivers and streams.

How much drinkable **freshwater** is there on Earth?

Only 2.59 percent of all the water on our planet is freshwater. However, much of that water is now polluted, and hydrologists and environmentalists estimate that only about one percent of the planet's total water supply is clean enough to drink.

What is **Arctic sea smoke**?

When extremely cold air blows over Arctic ice packs, the warmer seawater beneath causes fog to form when it comes into contact with the colder air. As the fog rises, it may appear to be smoke plumes.

What are **growlers**?

Growlers are pieces of floating ice that have broken off from an iceberg.

Which **freezes** more quickly— **cold or hot water**?

An old wives' tale that still circulates in American homes is that, if you wish to freeze ice quickly in an ice tray, put hot tap water in it and then put it in the freezer. Of course, many homes have automatic ice dispensers in their refrigerators, so this tale is beginning to die out somewhat. Let's dispel the myth now, however, and say that, no, hot water will not freeze faster than cold. However, what does work is boiling the water, then allowing it to cool down to a tepid temperature. Boiling the water removes air bubbles in the water, increasing thermal conductivity, and allow-

What unusual floating objects have been used to help chart currents in the oceans?

In a rather humorous example of kismet, oceanographers have been taking advantage of a 1990 accident in which a Korean cargo ship accidentally dumped 80,000 Nike shoes into the ocean. Since then, whenever these shoes have been found floating in the Pacific Ocean and elsewhere, oceanographers have taken note, tracking where the accident originally occurred and comparing it with the location where the shoes were found. In this way, they were able to gather additional information about currents.

It wasn't long before another accident created a new opportunity for studying the currents. In January 1992 a ship hauling toys lost part of its cargo in a storm. Nearly 30,000 rubber ducks, frogs, turtles, and beavers fell into the ocean. As with the athletic shoes, these toys, as they washed up on various shorelines, served as excellent indicators of the course taken by ocean currents.

ing the water to freeze more quickly. Of course, the time wasted boiling the water and then allowing it to cool would have been better spent by simply putting lukewarm water in the freezer in the first place! Not to mention that it would be more energy efficient not to boil the water.

Do **oceans** get more **rain** than land?

The oceans receive just over their share, percentage-wise, of the world's precipitation, about 70 percent. The remaining 30 percent of precipitation falls on the continents. Some areas of the world receive far more precipitation than others. Some parts of equatorial South America, Africa, Southeast Asia, and nearby islands receive over 200 inches (500 centimeters) of rain a year, while some desert areas receive only a fraction of an inch of rain per year.

What is **lake effect snow**?

Water evaporating off lakes can increase air humidity and encourage cloud formation and precipitation in areas near shorelines. As cold air moves over a water body such as a lake, it picks up moisture; then, when the clouds reach the shoreline, the air moves upwards in what is called the orographic effect. This, in turn, creates concentrated bands of precipitation that then release precipitation within relatively short distances of shorelines.

Some record snowfalls have been documented as a result of the lake effect. For example, on January 17, 1959, 51 inches (129.5 centimeters) of snow fell on Bennetts Bridge, New York, over a 16-hour period. Buffalo, New York, is infamous for being buried in snow because of its proximity to Lake Erie. The winter of 1976 to 1977, for instance, saw the city experience 30-foot snow drifts. Storms with 70-

mile- (113-kilometer)-per-hour winds and bitter temperatures led to the deaths of 29 people that season.

What is the **thermocline**?

Imagine the water in a sea or ocean as being similar to very dense air in the atmosphere. Just like with the air, layers of water can have different temperatures and pressures, with warm waters generally being on top of colder layers of water. The difference between the temperatures in these layers is called the thermocline.

At what temperature does **sea water freeze**?

While the amount of salt and other minerals and impurities in sea water makes a difference, in general sea water freezes at about 28°F (–2.2°C).

OCEAN CURRENTS

What are **ocean currents**?

The oceans don't remain still; their water is constantly moving in giant circles known as currents. In the Northern Hemisphere, surface currents move clockwise, while in the Southern Hemisphere they move counterclockwise. Currents help to moderate temperatures on land in places like the British Isles—which are farther north than the U.S.-Canadian border—by sending warm water from the Caribbean northeast across the Atlantic Ocean to northern Europe. A current known as the Antarctic Circumpolar Current circles the southern continent. The North Atlantic and North Pacific oceans each have a large clockwise current, while the South Atlantic and South Pacific oceans each have a large counterclockwise current.

What is an **Ekman spiral**?

Vagn Walfrid Ekman (1874–1954), a Swedish physicist and oceanographer, discovered that a combination of the Coriolis effect, the movement of surface waters, and the friction caused by winds blowing on the ocean's surface have an influence on current direction. In the Northern Hemisphere, the end result is that currents will be pushed to the right, and in the South the opposite occurs, forming spirals like weak whirlpools in the water. The effect, however, penetrates the ocean's surface only to a degree (the null point), as the influence of these factors weakens with water depth. (The layer affected by the spiral is called the Ekman layer). Scientists have noted that the Ekman spiral is most obvious underneath sea ice, because waves and other forces in the open sea nearly cancel out the effect. However, the Ekman spiral also applies to the Earth's atmosphere, where they are seen in surface winds.

How do **ocean currents affect weather**?

The world's oceans cover about 70 percent of our planet's surface. Thus, ocean water also absorbs more heat from the Sun than the land does. In addition to this, water absorbs and emits heat energy more slowly than land does. Warm water, therefore, remains warm longer, and cold water remains cold longer. As the world's ocean waters circulate through the action of currents, this warm or cold water can be transported long distances before it changes temperature. This is an important way that Earth distributes its heat energy. Warm waters in the tropical Atlantic Ocean are carried north as far as England and Scandinavia, for instance, while warm waters in the Indian Ocean circulate down toward Australia and South Africa. Land barriers, such as that formed by Central America, are vital for diverting currents in various directions. It's because of this that currents from western Africa move toward the Caribbean, and are then deflected north. Without Central America, Great Britain would be as frigidly cold as the remote wastes of northern Siberia.

Do **ocean currents** affect **coastal weather**?

Yes, ocean currents have the effect of moderating climates along coastal regions. Ocean waters tend to stabilize temperatures, which is why living on the coast of Southern California, for example, is a much more pleasant climatic experience than living farther inland, where it is much drier and warmer.

What are the **fastest currents** in the world's oceans?

The fastest known current is part of the Gulf Stream system: It heads north from the southern tip of Florida toward Cape Hatteras, North Carolina, flowing up to 3.3 to 6.6 feet (1 to 2 meters) per second, with some estimates as high as 7.5 feet (2.3 meters) per second. The Kuro Siwo Current in the Pacific Ocean runs a close second in terms of speed—around 1.3 to 4 feet (0.4 to 1.2 meters) per second.

Which **ocean currents** are the **slowest**?

The slowest currents are found deep in the world's oceans. These "sluggish," cold waters take up to about 1,000 years to circulate around the entire globe.

What is the **average temperature** of the world's oceans?

The average temperature of the oceans and seas is about 40°F (4.4°C).

What are the **major** surface and subsurface **currents** in the oceans?

The major warm ocean surface currents are the North Equatorial, Kuro Siwo, Gulf Stream, and South Equatorial Currents, plus the Equatorial Countercur-

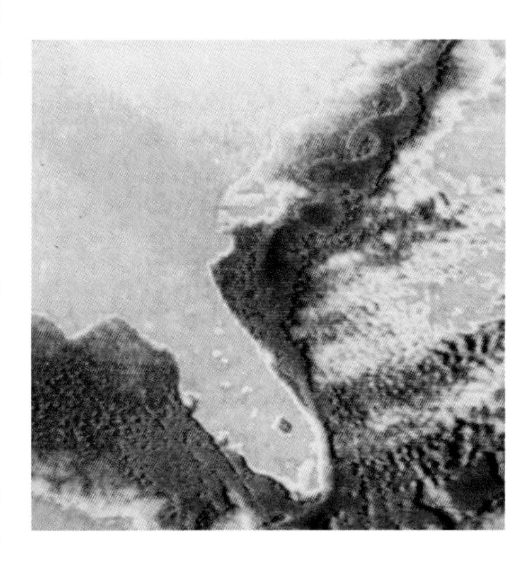

In this 1968 image, infrared cameras show the warmth of the Gulf Stream flowing off the U.S. East Coast. (*NOAA*)

rent and North Atlantic Drift. The major cold ocean surface currents are the Oyashio, California, Labrador, Peru (Humboldt), Benguela, Canaries, and Antarctic Circumpolar Currents. The major subsurface currents—all of them cold—are the Cromwell Current, Weddell Sea Bottom Water, Deep Western Boundary Current (the largest deep-ocean current), and the North Atlantic Deep Water.

What causes **deep ocean currents** to **flow** around the world?

The deep ocean currents are driven by thermohaline circulation, which is movement caused by differences in the temperature and salinity content of the water. Because cold, salt-laden water is heavier than warm water, it sinks to the bottom of oceans. To replace it, warmer water fills in, and as it subsequently cools, the rotation is repeated. This constant movement of water has often been referred to as a giant global conveyor belt or pump that slowly circulates water all over the world's oceans.

For example, the warm Gulf Stream current is heated by the Sun, "starting" in the Caribbean. It then flows north along the east coast of North America (mostly along the United States coastline) until it reaches sub-polar waters in the North Atlantic. Between Greenland and Norway, the cold Arctic winds cool the salt-laden water almost to the freezing point. Huge amounts of the now-cold, heavy salt water sink at this point to depths of around 3 to 4 miles (5 to 6.5 kilometers) and begin the next phase of the journey, traveling southwards through the Western Atlantic Basin to the Antarctic Circumpolar Current, and then into the Indian and Pacific Oceans. This trip takes many years. Off the coasts of Peru and California, for instance, upwellings often consist of ocean waters that sank to the depths centuries before.

How does the **Gulf Stream** affect the **weather** of the **British Isles** and **Scandinavia**?

The Gulf Stream currents move warm water in the Atlantic Ocean from the Caribbean Sea northeast to northern Europe and the British Isles. The current flows in this manner because currents heading west from the south Atlantic run into Central America. If North and South America were not connected by this relatively thin strip of land, the current would continue to flow westward and none of the warmer waters would reach Europe. Scotland, Ireland, England, Wales, and the Scandinavian countries would have climates more closely resembling Greenland's. When you look at a world map, you will notice that England, for instance, is at a latitude north of Newfoundland. Logically, it should be a lot colder there than it is, but thanks to the Gulf Stream, England has a climate more like that of New York City than northern Canada. Indeed, mainland Europe is warmer than North America because the Gulf Stream brings warmer weather with it.

What is **El Niño**?

El Niño refers to the phenomenon that occurs when warming waters in the tropical Pacific build up around Christmastime. It is a fairly regular occurrence (every two to seven years, but usually every three to four years, and with powerful ones

What is the story behind the man who lent his name to the Humboldt Current?

Next to Napoleon Bonaparte, Alexander von Humboldt (full name, Friedrich Wilhelm Heinrich Alexander Freiherr von Humboldt, 1769–1859) was largely regarded as the most famous man in Europe in his day. Educated in everything from finance and languages to astronomy, geology, and anatomy, Humboldt had a passion for science and travel. Most notable was his voyage to South America (1799–1804), during which he explored the natural landscape, making observations about the animal and plant life, as well as geological and astronomical observations. From this experience, he originated the idea that species vary depending on the climate, which varies with temperature and elevation. He also correctly surmised volcanoes probably align themselves along geological fissures in the Earth's crust. Humboldt connected the dots between geology and weather, noting how climate changes with elevation; he also was the first to observe how the Earth's magnetic field varies with latitude, and made observations that would later contribute to theories on how weather systems are generated with the planet's middle latitudes. When Humboldt's expedition reached the Pacific, he discovered the Peru Current, which is now also known as the Humboldt Current. For all his explorations and discoveries, Humboldt became a hero when he arrived back in Europe. He wrote about his journeys in the epic 30-volume work *The Voyage of Humboldt and Bonpland* (1805–1834) and was even more lauded for his two-volume *Cosmos* (1845, 1847), in which he attempted to unify many scientific disciplines to describe the complexity of nature as a whole system.

every 10 to 15 years), and because it happens around Christmas, Spanish-speaking people called it El Niño after the Christ child. Meteorologists often refer to it as the ENSO, which is short for El Niño Southern Oscillation. The term Southern Oscillation was coined by Sir Gilbert Thomas Walker (1868–1958), a British statistician and physicist who was studying the pattern of Indian monsoon seasons. The El Niño effect influences weather not only along the Pacific coasts of the Americas, but also worldwide. The consequences range from colder-than-normal temperatures in the central and eastern United States to strong storms in Africa, Australia, and the coast of California, flooding in Europe, and declines in fish populations in South America. Climatologists increasingly believe that strong El Niños increase droughts in Africa and may be accelerating desertification there. A moderate El Niño effect was in action during the years 2006 through 2008.

How does El Niño occur?

The El Niño weather phenomenon happens because of the unstable interaction of ocean surface temperatures with the lower atmosphere of the tropical Pacific Ocean in processes that are so complicated even modern science does not fully understand it. The process has to do with worldwide dynamics of the ocean and atmosphere, including oceanic waves crossing the entire planet, currents, and global circulation of the atmospheric. About the only thing that scientists have concluded with much certainty is that neither sunspots nor volcanic eruptions seem to influence El Niño.

A February 2007 image taken by the U.S.-French satellite Jason shows ocean temperatures (red for warmer, blue for cooler) as El Niño transitions to La Niña. (*NASA*)

What is **La Niña**?

It's pretty easy to guess that La Niña would be the opposite of El Niño. Instead of warm waters in the tropical Pacific, the surface waters are inordinately cool. In 2009, predictions were for a La Niña to dominate Pacific waters.

What years have been **El Niño years** versus **La Niña years** since 1950?

El Niño years since 1950 are: 1951, 1957-58, 1963, 1965, 1969, 1972, 1976-77, 1982-83, 1986-87, 1991-92, 1994-95, 1997-98 (the strongest in the past 50 years), 2002-03, 2004, 2006.

La Niña years include: 1950, 1954-56, 1962, 1964, 1968, 1970-71, 1973-76, 1984-85, 1988-89, 1995-96, 1998-2000, 2007-2009.

What is an **ENSO-neutral year**?

An ENSO-neutral year is a year during which neither El Niño nor La Niña conditions are present.

What are **atmospheric teleconnections**?

Atmospheric teleconnections refers to the way that conditions in the oceans, including such phenomena as El Niño, La Niña, and ocean currents, interact with and affect the weather.

WEATHER IN SPACE

Do **other planets** in our solar system experience **weather**?

Many planets and even moons in our solar system have atmospheres and, therefore, weather. The gas giants—Jupiter, Saturn, Uranus, and Neptune—all have very thick atmospheres which include storms and even lightning. Lightning has been observed on Jupiter, Saturn, Io, and Titan, and is speculated to exist on Venus. Jupiter, of course, is famous for the huge swirling "Great Red Spot," which is a wind storm 8,500 miles (14,000 kilometers) wide and 16,000 miles (26,000 kilometers) long that astronomers estimate has been ongoing for at least 400 years. Saturn, too, has a thick atmosphere topped by crystallized ammonia clouds; it also has a "Great White Spot," but this storm lasts for about a month and then fades, recurring at intervals. Uranus has a distinctive, blue-green atmosphere made mostly of hydrogen (83 percent) and helium (15 percent); Neptune also has a blue color (which is why it is named after the Roman god of the sea), with a hydrogen-helium-methane atmosphere. Below this upper atmosphere, astronomers believe Neptune has a lower atmosphere made of ammonia, crystallized methane, and ionized water; even deeper may be a layer of hydrogen sulfide. The winds on all of these planets are believed to be very powerful. Neptune, for instance, has winds reaching 700 miles (1,100 kilometers) per hour; the planet has a "Great Dark Spot," that is an ongoing storm about the size of Earth's Moon, and a smaller, white storm that has the whimsical name of "Scooter."

Not counting Earth, the inner planets—Mars, Venus, and Mercury—all have atmospheres. The atmosphere on Mars is very thin, and it has an air pressure about one percent that of Earth's. Theories abound as to why Mars lost its atmosphere, ranging from giant asteroids smashing into the planet and ripping away its air, to the lack of plate tectonics providing sufficient heat, to the idea that the planet somehow lost its magnetosphere. Without much atmosphere, Mars does not have much weather. Mercury, similarly, has virtually no atmosphere and is essentially a

The *Cassini* spacecraft took this image of Titan's atmosphere, with Saturn's south pole in the background, in December 2005. (*NASA/JPL/Space Science Institute*)

big rock in space. Venus is far more interesting than either Mars or Mercury, in terms of atmosphere. It is a planet with an atmosphere too rich in carbon dioxide, and consequently it experiences a "runaway greenhouse effect" that heats the surface to 900°F (500°C), which is hot enough to melt metals like lead, tin, and zinc.

Poor Pluto, which has been downgraded in status from planet to planetoid, has a thin atmosphere containing nitrogen, methane, and carbon monoxide. Several moons in our solar system also have atmospheres. Of these, Saturn's Titan is perhaps the most interesting. Its dense atmosphere (50 percent denser than Earth's) contains 98.4 percent nitrogen and 1.6 percent methane. Astronomers believe that Titan has lakes and seas composed of liquid methane and nitrogen, and that it experiences rain showers made of methane. A smoggy, orange haze made of hydrocarbons envelopes the atmosphere. Some other moons have thin atmospheres, such as Neptune's Triton, and Ganymede, which is another moon circling Jupiter, is believed to have a water ocean beneath a layer of ice.

Is there **weather outside the solar system**?

In recent years there has been a flurry of activity among astronomers who are discovering more and more exoplanets. Over 200 have thus far been found, and most of these are gas giants. The reason for this is because gas giants, being very large, exert a greater gravitational pull. The most successful technique for finding exoplanets has been to observe how stars move; observatories such as the Hubble Telescope are sensitive enough to detect if a star has a slight wobble to it, which indicates the presence of planets and a solar system. Combining this method with infrared spectroscopy, it is possible to detect the molecular composition of distant planets. Carbon dioxide and monoxide, hydrogen, neon gas, water, and hydrocarbons have all been detected. Astronomers are quite confident that they are seeing planets outside our solar system that have atmospheres, and they speculate that some may have life-supporting liquids and gasses.

THE MOON

What is **syzygy**?

Syzygy occurs when the Moon and Sun are aligned, an event that happens twice monthly. The effect on Earth is that the gravitational pull of these two space bodies is strengthened, increasing the elevation of high tides and decreasing low tides.

A total solar eclipse as seen from Baja California on July 11, 1991. *(NASA)*

What is the difference between a **lunar eclipse** and a **solar eclipse**?

A solar eclipse occurs when the Moon passes between the Earth and the Sun, while a lunar eclipse happens when the Earth is between the Moon and the Sun. In a solar eclipse, the Moon casts a shadow on the Earth's surface, while the opposite is true of a lunar eclipse.

What is a **blue moon**?

"Once in a blue moon" is an expression referring to an event that does not happen very often. A blue moon happens about once every 2.7 years when the Moon is full twice during the same calendar month. The exact origin of the phrase is not certain, but it seems to be about 400 years old and comes from English roots. To say that the Moon was blue would be to express that something was absurd or unlikely. Thus, the phrase did not mean that the Moon was literally blue.

Sometimes, however, the Moon has been known to take on a bluish hue, especially when there is a lot of ash or dust in the atmosphere, such as after a large forest fire or volcanic eruption.

What are **tides**?

Tides are the consequence of any two objects that exert gravitational pull on one another over a long period of time. Basically, each object gently pulls the other object into an egg-like shape, because the gravitational acceleration on one side of the object is larger than on the other side. On Earth, the most observable evidence of this gravitational effect is the changing tides we witness.

How do **tides work**?

Two cycles of high and low tides occur each day, roughly 13 hours apart. High tides occur both where the water is closest to the Moon, and where it is farthest away. At the points in between, there are low tides.

> **Does the air have tides?**
>
> Strangely enough, yes. Just as with water, the atmosphere can be affected by the Moon's tug on the Earth, so that air pressure can change in air masses daily. The changes are very small, though—about one or two millibars—and these changes are mostly seen only in equatorial zones.

How **often** do ocean **tides occur** on Earth?

During a 26-hour period, each point on Earth's surface moves through a series of two high tides and two low tides—first high, then low, then high again, then low again. The length of the cycle is the sum of Earth's period of rotation, or the length of its day (24 hours), and the Moon's eastward orbital movement around Earth (two hours).

Do **lakes** have **tides**?

Compared to oceans and seas, lakes generally do not contain enough water to have noticeable tides due to the gravitational effects of the Moon on the Earth. They do have tides, though, and large lakes, such as Lake Superior in North America, are big enough to have observable tides. Lake Superior's waters fluctuate about three inches.

Can **tides influence weather**?

Many scientists believe that tides have a small influence on the weather. Tides can have some effect upon ocean currents, for example, which, in turn, influence weather patterns. It is also known that the Moon's tug on our planet can actually cause measurable changes in topology (the Earth's crust actually moving up and down). This can, in turn, encourage earthquakes or volcano eruptions, and the latter definitely affects our weather. Tides, it can be said, may also be relevant to weather because, when strong onshore winds coincide with high tides, coastal flooding is made worse than would otherwise occur at low tides.

THE SUN

Why is the **Sun essential for weather**?

Our Sun is the source of almost all of the energy that creates weather on the planet; plate tectonics, the heat within the Earth, tidal effects, radioactive elements, and the effects of gravity from Jupiter and Saturn provides the rest. The majority of the energy—99.98 percent—is from the Sun, however. Although the amount of light, heat, and other energy from the Sun that hits the Earth is equal to only 5×10^{-10} of its total output, this is still the equivalent of burning 700 billion tons of coal every day! It has also been calculated that the energy reaching Earth's upper atmosphere is equal to 5 million horsepower *per square mile*. That's enough to rev up the engines of 7,400 Indianapolis 500 race cars. Planet-wide, that would be about 14.5 billion race cars being powered up daily by the Sun.

If the Sun's energy on the planet's surface was constant, it would not affect weather. This is not the case. Because our planet spins, giving us day and night, and tilts, giving us the seasons, the energy is not distributed evenly. Also, clouds and variations in the upper atmosphere affect solar energy levels. The result is that Earth is covered with areas of colder and warmer air, causing air masses to move, which are, in turn, diverted this way and that by variations in geography and the Coriolis effect. This is what gives us the weather.

How can a **solar eclipse** influence the **weather**?

During a solar eclipse, the Moon casts a dark shadow, called the umbra, onto the Earth's surface, as well as a lighter shadow called the penumbra. The umbra can be about 170 miles (274 kilometers) in diameter, darkening the sky and cooling the air just as if the Sun were setting. Temperatures have been known to drop many degrees during a solar eclipse, especially during a total eclipse. For example, in Baja, California, the temperature dropped from 90°F to 74°F (32°C to 23°C) on July 11, 1991, because of an eclipse. The effect, however, is short lived, as total eclipses rarely last more than seven minutes, and specific locations on the planet only experience them once every four centuries or so.

How much of the **Sun's energy** is **radiated back into space** by the Earth?

As the Sun provides us with energy, it is also being reflected back into space. Because this give and take is in balance, our planet stays at pretty much the same total temperature, which is a very good thing. If Earth reflected back *more* energy than it received, the planet would cool, eventually becoming an ice ball. If the opposite happened, as some scientists fear could be occurring with global warming, the planet would heat up. Interestingly, some planets, including Jupiter and Neptune, radiate much more heat than they receive, which has led to speculation that there is a heat source within these planets. Some astronomers consider Jupiter a kind of failed star; if it had been more massive, it could have become a second Sun, and Earth would be part of a binary star system.

What is a **blue Sun**?

An even rarer event than a blue Moon is a blue Sun. As with a real blue Moon, the Sun can appear bluish as a result of fires, volcanic eruptions, or sandstorms coating the air in dust, ash, or sand.

Does the **Sun** have an **atmosphere**?

Yes, in a manner of speaking. Since the Sun is made primarily out of hydrogen (73 percent) and helium (25 percent) in a plasma state, there's no visible solid surface to distinguish a dividing line between the Sun's body and its atmosphere. The "surface" of the Sun is a layer called the photosphere that is about 300 miles (480 kilo-

What is UVA and UVB radiation?

UVA and UVB are two categories within the ultraviolet range. UVA radiation has a wavelength between 320 and 400 nanometers, and UVB has a range of 290 to 320 nanometers. The more energetic UVB radiation is considered more dangerous to people, as it can lead to malignant melanomas. UVA can also potentially cause cancer, but is more likely to just cause unsightly wrinkles and leathery skin. This is why dermatologists and other physicians always recommend using sunscreen to prevent damage from ultraviolet radiation.

meters) thick and has an average temperature of about 10,000°F (5,500°C); above that is a second layer, called the chromosphere, that is thousands of miles thick, and is somewhat cooler—7,800°F (4,300°C). Above that is the Sun's corona, which is where temperatures kick into high gear. Temperatures in the corona average about 1.8 million °F (1 million °C).

How much **time** does it take for **sunlight** to reach the **Earth**?

The speed of light is about 186,000 miles (300,000 kilometers) per second, and the Sun is about 92.58 million miles (149 million kilometers) away, so it takes sunlight 8.4 minutes to reach our planet. Therefore, the Sun we see in the sky is actually as the star appeared over eight minutes ago! Not to paint too depressing a picture, but if the Sun went nova, no one on Earth would realize the end was near for eight and a half minutes.

What is the **solar spectrum**?

The solar spectrum is the spectrum of light that includes wavelengths both visible and invisible to the human eye. When diverted through a prism, this white light separates into the familiar rainbow spectrum of colors, ranging from violet to red. Temperature is affected by the longer-wavelength spectrum (red through invisible infrared). About 50 percent of the Sun's spectrum emerges as wavelengths at the infrared level, and 10 percent are in the ultraviolet range.

Why is **infrared light** from the Sun **important**?

Infrared light from the Sun provides about 50 percent of the solar energy that reaches Earth in the form of heat. Without it, our planet would be a much colder place. The rest of the heat energy comes from light in the visible spectrum range that, after it is absorbed, radiates back into the atmosphere in the form of heat.

How is the **time set** for the beginning of **sunrise and sunset**?

In the United States, sunrise officially begins when the leading edge of the Sun disc first begins to rise over the horizon; and sunset happens when the top of the

Sun disappears behind the horizon. In Great Britain, however, sunrise and sunset are measured by the time when the middle of the Sun is at the edge of the horizon. Scientists use highly sensitive light-measuring instruments to calculate when dawn is approaching as much as 90 minutes before sunrise actually occurs.

What is a **pyranometer**?

A pyranometer (also called a solarimeter) is an instrument that measures the amount of sunshine reaching the ground. Pyranometer comes from the Greek words "pyr," meaning fire, and "ano," meaning sky. The instruments record the amount of sunlight in terms of watts per square meter, and they calculate this in one of two ways. A less accurate method used in cheaper models of pyranometers detects light using small, silicon-based photodetectors. This method is not considered to be ideal because such photodetectors do not capture the full spectrum of sunlight very well. The other type of pyranometer uses a thermopile, which is a collection of highly sensitive thermocouples that detect temperature changes across junctions. Pyranometers can be used to evaluate weather conditions and, when readings are taken over extended periods of time, changes in climate.

This Angstrom pyranometer from 1930 measured albedo, which is the amount of electromagnetic radiation reflected from the planet's surface. *(NOAA)*

What is a **Campbell-Stokes recorder**?

A Campbell-Stokes recorder is a device for measuring not only the amount of sunlight reaching the ground, but also its intensity. Also called a Stokes sphere, it was invented in 1853 by John Francis Campbell (1821–1885), a Scotsman who was actually a Celtic scholar; his name is also cited as Iain Frangan Caimbeul an Iain Òg Ìle. Using a glass sphere to focus the Sun's rays, Campbell placed a card with different marks on it to indicate times. Depending on the light's intensity, the energy magnified by the glass burns the card. Differently marked cards are used depending on the season and whether one is in the Northern or Southern Hemisphere. The name Stokes comes from the English mathematician and physicist Sir George Gabriel Stokes (1819–1903), who improved the casing (he used metal) and changed the arrangement of the glass sphere and cards.

What type of **pyranometer** does the **National Weather Service** use?

Instead of the Campbell-Stokes recorder, the NWS uses a device called the Marvin sunshine recorder. A more elaborate device, it consists of one clear and one blackened bulb connected by a thin glass tube partly filled with mercury. As sunlight warms the blackened bulb more than the clear one, the mercury expands and shorts out an electrical contract that then operates a pen that marks a chronograph.

Does the Sun **emit a constant amount of energy**?

The energy the Sun emits—technically called "solar irradiance"—currently is much more than it did when the solar system was young. Scientists estimate that, billions of years ago, our Sun only emitted about 75 percent of the energy it now does. Now, however, the amount of solar irradiance is fairly stable, though not completely constant. It can vary by as much as 0.1 percent, which may not sound like much, but such fluctuations are big enough to make the difference between moderate and frigid winters, merely warm or searing-hot summers. Solar irradiance can be affected by sunspots, but the internal workings of the Sun are still so complex that scientists do not fully understand what causes these fluctuations. Despite there being no complete consistency in the Sun's energy output, astronomers and meteorologists still refer to a "solar constant," which is the amount of energy of all emissions (measured in watts) per square meter. Readings are taken at the upper extremes of the Earth's atmosphere, which receives, on average, about 1,366 W/m^2.

What **U.S. cities** receive the most **sunlight** annually?

Yuma, Arizona, is regarded as the sunniest city in the United States, with about 4,000 hours of sunlight every year. Below are some other sunny cities for those who are averse to rainfall. Because what constitutes a "sunny day" is debatable, the figures are estimates of the percentage of hours of annual sunshine, or what meteorologists call the "mean percentage of possible sunshine."

Sunniest U.S. Cities

City	Percent of Annual Sunshine
Yuma, AZ	90%
Redding, CA	88%
Las Vegas, NV	85%
Tucson, AZ	85%
Phoenix, AZ	85%
El Paso, TX	84%
Fresno, CA	79%
Reno, NV	79%
Bishop, CA	78%
Santa Barbara, CA	78%
Sacramento, CA	78%
Bakersfield, CA	78%
Flagstaff, AZ	78%
Albuquerque, NM	76%

Which state is known as the Sunshine State?

Florida has adopted the name of Sunshine State. Although it does have a warm, often sunny climate, it also has a pronounced rainy season from April through October. Ironically, Florida actually has more thunderstorms than any other U.S. state. Arizona actually deserves the credit of being the "true" sunshine state.

What is a person's **circadian rhythm**?

Human beings have evolved in a world where there are regular cycles of day and night, and we have thus developed a kind of internal clock that makes us want to sleep at night and be awake during the day. This occurs even when we are, say, underground or in an enclosed factory with no windows where there are no visual cues as to the time of day. The circadian rhythm controls such bodily functions as peristalsis (movement of food through the digestive tract, leading eventually to excretion), blood pressure, melatonin secretion, hormone levels, and alertness versus fatigue. In our modern world, people have tried to adjust to lifestyles that are not exactly commensurate with their circadian rhythms. Electricity and other sources of power allow us to work late at night, drive vehicles, and operate machinery at times when our bodies would rather be asleep. The result is that more accidents involving cars and factory equipment happen at night than during the day. We are also much less efficient at night, even when doing safe activities like balancing the checkbook.

Are some people **allergic to the Sun**?

There are two rather odd reactions some people have to the Sun. One, allergic conjunctivitis, is relatively harmless. This is when sudden exposure to the Sun, such as when leaving a darkened building, causes a sneezing reaction inspired by the Sun's ultraviolet radiation. It can also cause redness and itchiness in the eyes, which is easily remedied with sunglasses and antihistamines. More serious is something called "Sun poisoning," which causes a rash known as polymorphous light eruption (PLE), with itchy bumps and swelling of the skin. Even suntan lotion does not prevent this reaction for people who are ultra-sensitive to the Sun's rays. Topical corticosteroids are typically prescribed by doctors to treat the rash.

What is the **sunburn index**?

The sunburn index—which is also known as the UV index—was developed by the National Weather Service and the Environmental Protection Agency as a way to provide people with information as to the varying risks of sunburn on particular days. Ranging in scale from 0 to 10 (10 being the riskiest), the index is calculated based on several factors: day of the year, latitude, elevation, cloud cover, and ozone levels. People with fair skin should be the most cognizant of the UV index when ven-

UV radiation can damage the skin within minutes, depending on exposure. A strong sunscreen with a high SPF is recommended by most doctors for this reason.

turing outside, though people with tan or dark complexions should not ignore it either. The table below explains the index in more detail.

Sunburn (UV) Index

Index	Exposure Level	Estimated Minutes before Skin Damage	Recommendations
1-2	Minimal	30-120	Wear a wide-brimmed hat
3-4	Low	15-90	Wear hat, sunglasses, long-sleeve shirt or other covering
5-6	Moderate	10-60	Wear hat, sunglasses, cover up with clothes, seek shade
7-8	High	7-40	Wear hat, sunglasses, cover up with clothes, seek shade, limit exposure to sunlight
9-10	Very High	3-30	All of the above while minimizing exposure to the Sun, or stay indoors

What else does **too much Sun** do to one's skin?

Besides increasing the risk of skin cancer (melanoma), Sun exposure causes premature wrinkling of the skin and eye cataracts.

What is the SPF system?

The Sun protection factor (or SPF) rates sunscreens based on how much they shield your skin from ultraviolet radiation. Depending on your skin pigmentation and age, an SPF rating of somewhere between 30 and 50 is usually recommended for any lengthy amount of time in the Sun, and children, especially, should use higher SPF-rated sunscreens. Sunscreen should be applied about a half hour before going outside. If you are unsure about what sunscreen to use, a good idea is to ask your physician or pharmacist.

Why is the **sunniest place** on Earth also **the coldest**?

Because of the Earth's tilt, Antarctica receives more rays from the Sun than any other land mass on the planet. However, because the continent is covered in snow, the white stuff reflects much of that energy back into the atmosphere—50 to 90 percent of it, in fact. Thus, Antarctica is extremely cold. This was not always the case, though. Millions of years ago, during the Jurassic Period, Antarctica was much closer to the equator. Indeed, scientists are discovering dinosaur fossils there.

Can people's **moods** be affected by **sunlight**?

It is well known that, during the winter, people living in the Northern Hemisphere tend to experience more depression and listlessness. Sometimes called Seasonal Affective Disorder (SAD), this syndrome may be caused by chemical changes in the brain during extended periods of darkness. In addition, the human body needs at least a few minutes of sunlight a day to generate Vitamin D. Vitamin D deficiency is a chronic problem in many northern states in America, where it can lead to feelings of depression and a lack of energy and even libido.

What is **albedo**?

Albedo is the amount of the Sun's energy that is reflected back from the surface of the Earth. Overall, about 33 percent of the Sun's energy bounces off the Earth and its atmosphere and travels back into space. Albedo is usually expressed as a percentage.

How much **sunlight** is **reflected by snow**?

Since white reflects snow, it should be no surprise that snow cover bounces sunlight back into the atmosphere and into space at an efficient rate of about 80 percent. Environmentalists speculate that global warming may be accelerated further because, as snow melts, more and more of the Sun's energy will be absorbed by the ground.

SUNSPOTS AND SOLAR ACTIVITY

What is a **sunspot**?

Sunspots, when viewed by visible light, appear as dark blemishes on the Sun. Most sunspots have two physical components: the umbra, which is a smaller, dark, featureless core, and the penumbra, which is a large, lighter surrounding region. Within the penumbra are delicate-looking filaments that extend outward like spokes on a bicycle wheel. Sunspots vary in size and tend to be clustered in groups; many of them far exceed the size of our planet and could easily swallow Earth whole.

Sunspots are the sites of incredibly powerful, magnetically driven phenomena. Even though they look calm and quiet in visible light, pictures of sunspots taken in ultraviolet light and in X-rays clearly show the tremendous energy they produce and release, as well as the powerful magnetic fields that permeate and surround them.

Who first **discovered sunspots**?

The earliest recorded observations of sunspots go all the way back to the year 28 B.C.E., when Chinese astronomers made note of dark spots on the Sun. In the era of modern, western civilization, the credit goes to the famous Galileo Galilei (1564–1642), who first recognized sunspot activity through his telescope in around 1611 (sources vary, crediting the discovering anywhere between 1610 and 1613). Records show that others, including Johannes Kepler (1571–1630), had observed sunspots before Galileo, but they failed to recognize them for what they were. Kepler, for instance, mistook the spot he saw, several years before Galileo, as the planet Mercury orbiting in front of the Sun.

How **big** do **sunspots** get?

Sunspots can range from the relatively small and, except with a telescope, unobservable, to the staggeringly enormous. The biggest sunspots can be more than 100,000 miles (161,000 kilometers) across. Astronomers measure sunspots in "millionths," with each millionth being one millionth of the surface area of the Sun that is facing the Earth. The Earth, if it were a sunspot on the surface of the Sun, would be equal to 169 millionths. Compare that with the typical sunspot, which ranges from 300 to 500 millionths, and you get an idea of how big they are. One of the largest sunspots ever measured was seen in 2001 and was 2,400 millionths. But that does not account for how far the solar flares emanating from sunspots shoot out into space; solar flares can be as long as 100,000 miles (161 thousand kilometers), and some of the energy they emit can literally stretch to Earth's orbit, 93 million miles (150 million kilometers) away.

Is there a **sunspot cycle**?

Yes. Sunspot activity goes through a two-cycle pattern of high and low activity: one that lasts about 11 years (more accurately, about 10 years, six months), and also an

The Sun's chromosphere is captured in this photo by Hinode's Solar Optical Telescope in 2007. The Hinode mission is funded by the United States, Japan, the United Kingdom, and the European Space Agency. *(JAXA/NASA)*

88-year cycle of highs and lows; and astronomers speculate an even longer pattern might be possible. The first person to observe sunspot cycles was German astronomer Heinrich Samuel Schwabe (1789–1875). Originally a pharmacist, Schwabe became an amateur and then professional astronomer. Wondering whether there was another planet besides Mercury and Venus close to the Sun, he accidentally discovered sunspots and became completely fascinated by them. From 1825 until near the end of his life, he observed the Sun daily, recording the number of sunspots he observed. From these observations, he noticed that there were periods of greater and lesser activity, which he thought came in cycles of about 10 years, a fairly close approximation, given the quality of telescopes at the time.

What is the **Maunder Minimum**?

Named after English astronomer Edward W. Maunder (1851–1928), the Maunder Minimum was a period of extremely low sunspot activity that lasted from 1645 to 1715. Maunder discovered this solar event by researching old records.

What are some other **sunspot minimums and maximums**?

Below is a list of notable sunspot activity changes dating back to 1000 C.E. Some of the names for these maximums and minimums come from their discoverers,

Why do sunspots appear dark?

Sunspots are a slightly cooler temperature (about 2,000 degrees Fahrenheit [1,100 degrees Celsius] cooler) than their surrounding photospheric gas, and so in the bright back-lighting, sunspots appear dark. Do not be fooled, though; a sunspot is still many thousands of degrees, and the amount of electromagnetic energy that courses through sunspots is tremendous.

including German astronomer Gustav Spörer (1822–1895), English meteorologist John Dalton (1766–1844), English astronomer Edward W. Maunder (1851–1928), and Swiss astronomer Johann Rudolf Wolf (1816–1893).

Sunspot Activity 1000 C.E. to Present

Years (c.e.)	Name
1010 to 1050	Oort Minimum
1100 to 1250	Medieval Maximum
1280 to 1340	Wolf Minimum
1420 to 1530	Spörer Minimum
1645 to 1715	Maunder Minimum
1790 to 1820	Dalton Minimum
1950 to Present	Modern Maximum

What is the **Wolf Number**?

Swiss astronomer Johann Rudolf Wolf (1816–1893) devised a system for counting sunspots in 1848. The Wolf Number is named in his honor. Astronomers count sunspots from various observation points across the globe, then average those counts to come up with the official Wolf Number for that period.

What happens to the **Sun every 22 years**?

The sunspot activity on the Sun is related to the magnetic field change the Sun goes through. Every 22 years, the Sun's magnetic field completely reverses. Magnetic north becomes magnetic south and vice versa. Other than having something to do with sunspot activity, astronomers do not believe that this flipping affects our weather in any way.

What is a **solar prominence**?

Prominences are high-density streams of solar gas projecting outward from the Sun's surface (photosphere) into the inner part of the corona. They can be more than 100,000 miles long and can maintain their shapes for days, weeks, or even months before breaking down.

What were the Great Solar Storms of 1989?

In March 1989 solar activity became much more active than usual, creating solar storms that sent out high energy particles into the Earth's ionosphere. The storms had adverse effects on communications and weather satellites, and also resulted in spectacular aurorae that reached as far south as Mexico. The storms also affected power grids in some areas. Most notably, six million people in Quebec, Canada, experienced a blackout when circuit breakers and fuses overloaded. More recently, an even bigger solar storm occurred on Halloween 2003, mostly affecting Sweden with power outages and damaging 28 satellites, two of which were complete losses.

What is a **coronal mass ejection**?

A coronal mass ejection is a huge blob of solar material—usually highly energetic plasma—that is thrown outward into space in a huge solar surface explosion. Coronal mass ejections are associated with solar flares, but the two phenomena do not always occur together. When coronal mass ejections reach the space near Earth, artificial satellites can be damaged by the sudden electromagnetic surge caused by the flux of these charged particles.

What is a **solar flare**?

Solar flares are sudden, powerful explosions on the surface of the Sun. They usually occur when large, powerful sunspots have their magnetic fields too tightly twisted and torqued by the hot, swirling plasma in the Sun. The magnetic field lines unwind and break suddenly, and the matter and energy that had been contained rushes outward from the Sun. Solar flares can be many thousands of miles long, and they can contain far more energy than all of the energy consumption of all of human history on Earth.

What is **solar wind**?

Our Sun is constantly emitting high-energy matter in the form of electrons, protons, and other particles. This "solar wind," as it is called, is largely deflected by the Earth's magnetic field, sometimes penetrating the upper atmosphere. While there is no scientific proof, some scientists believe that changes in the solar wind can have long-term effects on our planet's climate.

What is a **geomagnetic storm**?

A geomagnetic storm is the term used for a strong increase in solar wind activity—including intense X-rays—that affects the Earth's magnetic field.

What effect does solar activity have on life here on Earth?

By the time the solar wind reaches the distance of Earth's orbit, its density is only a handful of particles per cubic inch. Even so, it is enough to have caused substantial radiation damage to life on Earth over the several billion years of Earth's history, if not for Earth's protective magnetosphere.

When solar activity is particularly strong, such as during a solar flare, the stream of charged particles can increase dramatically. In that case, these ions can strike molecules in the upper atmosphere, causing them to glow. Those eerie, shimmering lights are called the Aurora Borealis (Northern Lights) and Aurora Australis (Southern Lights). During this time, Earth's magnetic field can temporarily weaken, causing our atmosphere to expand; this can affect the motion of satellites in high-Earth orbit. In extremely strong periods of solar flux, electrical power grids can be affected.

How **fast** does the **solar wind** travel?

The flow of plasma out from the Sun is generally continuous in all directions, typically moving at speeds of several hundred kilometers per second. It can, however, gust out of holes in the solar corona at 2,200,000 miles per hour (1,000 kilometers per second) or faster. As the solar wind travels farther from the Sun, it picks up speed, but it also rapidly loses density.

How **far** does the **solar wind** travel?

The Sun's corona extends millions of miles beyond the Sun's surface. The plasma of the solar wind, however, extends billions of miles farther—well beyond the orbit of Pluto. Beyond there, the plasma density continues to drop. There is a limit, called the heliopause, where the influence of the solar wind dwindles to just about nothing. The region inside the heliopause—which is thought to be some 8 to 14 billion miles (13 to 22 billion kilometers) from the Sun—is called the heliosphere.

How do **sunspots affect weather**?

While scientists still debate how important or significant sunspot activity is on our weather and climate, there are some theories. In the short term, solar flares and winds do not influence weather to any significant extent because the magnetosphere and upper atmosphere are more than capable of absorbing their energy. Over the long term, however, there could be consequences if the Sun experienced a prolonged or permanent change in activity.

Fluctuations in solar activity typically occur in the ultraviolet wavelengths, and UV radiation is known to affect the Earth's upper atmosphere. Astronomers and some meteorologists speculate that X-rays, too, resulting from sunspots and flares could, over time, change the amount of nitric oxide (NO) in the upper atmosphere, and this would have an effect on the ozone. Sunspot activity can result in more cosmic rays penetrating the atmosphere, which, in turn, spur on cloud formation and increase precipitation.

More recently, there have been theories that significant decreases in solar activity predict oncoming ice ages on our planet. For example, during the Little Ice Age that lasted from the fifteenth through the eighteenth centuries (a period including the Maunder Minimum), sunspot activity was at a low point. Other minimums (the Dalton Minimum [1790—1820] and Spörer Minimum [1420–1530]) also coincide with colder weather.

A photo taken at the Dryden Flight Research Center in Edwards, California, took advantage of a smoky sky to reveal sun spots. *(photo by Tom Tschida courtesy NASA)*

What has **recent sunspot activity** been doing?

Sunspot activity during the dawning years of the twenty-first century has been a rollercoaster ride. In 2004, it was reported that there were more sunspots on the Sun than the Earth had seen in some 8,000 years. The next year, however, there was an abrupt drop-off in sunspot activity, a trend that has continued through 2008 and into 2009. In fact, during the last months of 2008, there were almost no sunspots at all.

Is it possible to **forecast space weather**?

By understanding past solar activity patterns and comparing them to current observations, astronomers can make some rough predictions about possible upcoming geomagnetic storms. This type of forecasting is important, so that warnings may be issued concerning dangers to orbiting satellites and other space missions, as well as the risks posed to electric power grids on Earth.

Who managed to **calculate cosmic ray activity on Earth** dating back over 11,000 years?

Sami Solanki (1958–), an astronomy professor at Germany's Max Planck Institute, developed a method for estimating solar ray activity dating back millennia. Solanki based his method on the knowledge that cosmic rays create chemical reactions in the atmosphere. One byproduct of these reactions is carbon-14, which then precipitates down onto the Earth's crust. Trees and other vegetation absorb this radioactive form of carbon; preserved vegetation that has been buried underground for centuries can be dug up and its composition analyzed. Radiation from sunspots actually decreases the production of carbon-14, and so when sunspot activity is high, there is less carbon-14 for trees to absorb. Using this information, Solanki discovered that there has been more sunspot activity since around 1930 than during any other period going back 8,000 years.

THE MAGNETIC FIELD

What is **Earth's magnetic field**?

Electromagnetic force permeates our planet. In essence, Earth itself acts like a giant spherical magnet. This is caused primarily by the motion of electrical currents within Earth, probably through the liquid metallic part of Earth's core. Combined with Earth's rotation, the core acts like an electric dynamo, or generator, creating a magnetic field.

Earth's magnetic field extends thousands of miles outward into space. Magnetic field lines, carrying and projecting electromagnetic force, anchor at Earth's magnetic poles (north and south) and bulge outward, usually in large loops. Occasionally, they stream outward into space. The magnetic north and magnetic south poles of Earth's magnetic field are very close to the geographical north and south poles, which mark the axis of Earth's rotation. (Be careful, by the way. There are two ways to define Earth's magnetic poles—the "magnetic north pole" is on an island in Canada, but the "geomagnetic north pole" is actually on Greenland, and the "geographic north pole" is on an ice shelf floating on the ocean, hundreds of miles from any land.)

How did people **discover** that Earth has a **magnetic field**?

The ancient Chinese were the first to use magnets as compasses for navigation. Though they did not know it, these "south-pointing needles" worked because the magnets aligned themselves with Earth's magnetic field. Since Earth's magnetic poles have been very close to the rotational north and south poles, compasses point almost exactly north and south in most parts of the world.

Over time, scientists started making a connection between lodestones (permanent magnets) and the nature of Earth itself. The English astronomer Edmund Halley (1656–1742), for example, spent two years crossing the Atlantic on a Royal Navy ship, studying Earth's magnetic field. Later, the German mathematician and scientist Karl Friedrich Gauss (1777–1855) made important discoveries about how magnets and magnetic fields work in general. He also created the first specialized observatory for the study of Earth's magnetic field. With his colleague Wilhelm Weber (1804–1891), who was also famous for his work with electricity, Gauss calculated the location of Earth's magnetic poles. (Today, a unit of magnetic field strength is called a gauss in his honor.)

Why is Earth's **magnetic field important** to life on Earth?

Earth's magnetic field extends out into space, creating a structure called a magnetosphere, which surrounds our planet. When the magnetosphere is hit by charged

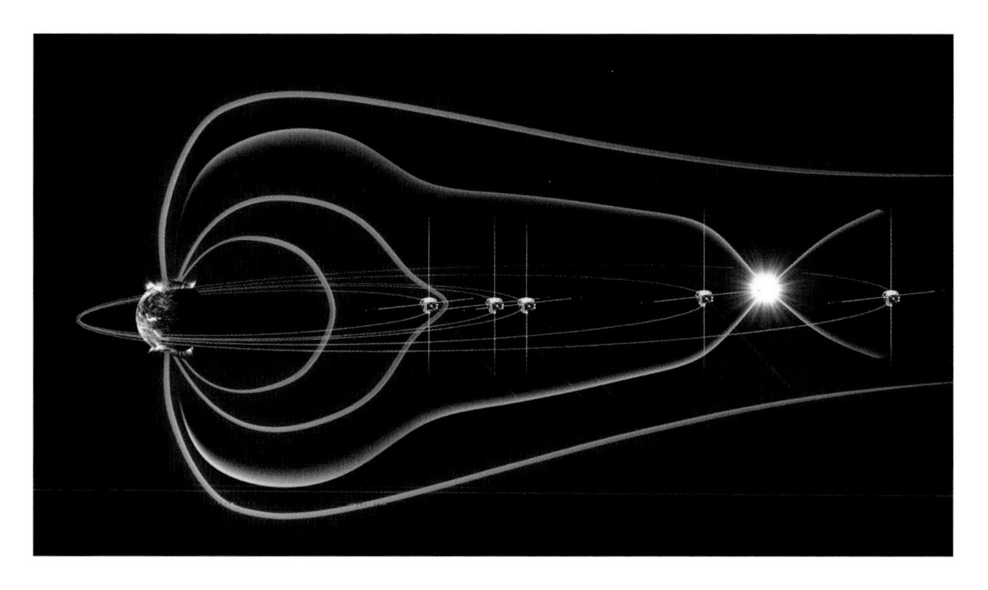

A graphic from NASA depicts the THEMIS (Time History of Events and Macroscale Interactions during Substorms) mission to study the Earth's magnetic fields (blue lines) and substorms (aurora that result from intense space storms). THEMIS's orbit changes to varying degrees in order to better pinpoint the location of substorms. (*NASA*)

particles from space, such as from the solar wind or from a coronal mass ejection, it deflects these particles away from Earth's surface, significantly reducing the amount that strikes life forms down on Earth's surface. This protects us from the hazards of being hit by too many such particles.

What role did **Walter Maurice Elsasser** play in discovering the magnetic field?

German-born American physicist Walter Maurice Elsasser (1904–1991) was the scientist who discovered that the Earth's magnetic field is generated as a result of how the planet's hot core acts like a dynamo. He also brilliantly discovered that analysis of rock particles reveal clues about the orientation of the Earth's magnetic field over time, which in turn contributed to the science of plate tectonics and the history of Earth's climate.

How does the **magnetosphere** help **animal behavior** on Earth?

Earth's magnetic field is also very important to animals on Earth that migrate or otherwise travel long distances. Some animals have impressive built-in magnetic sensors; biologists have shown that many migratory birds figure out where to fly by using Earth's magnetic field to guide them. Humans, too, benefit from the magnetosphere by using magnetic compasses to figure out which way is north or south.

How **strong** is Earth's **magnetic field**?

On typical human scales, it is pretty weak; at Earth's surface, it is about one gauss in most places. (A refrigerator magnet is typically 10 to 100 gauss.) However, the

How do we know Earth's magnetic field can flip upside down?

In 1906 French physicist Bernard Brunhes (1867–1910) found rocks with magnetic fields oriented opposite to that of Earth's magnetic field. He proposed that those rocks had been laid down at a time when Earth's magnetic field was oriented opposite to the way it is today. Brunhes's idea received support from the research of Japanese geophysicist Motonori Matuyama (1884–1958), who in 1929 studied ancient rocks and determined that Earth's magnetic field had flipped its orientation a number of times over the history of our planet. Today, studies of both rock and the fossilized microorganisms imbedded in the rock show that at least nine reversals of Earth's magnetic field orientation have occurred over the past 3.6 million years.

The exact cause of the polarity reversal of Earth's magnetic field is still unknown. Current hypotheses suggest that the reversal is caused by Earth's internal processes, rather than external influences like solar activity.

energy in a magnetic field depends strongly on its volume; so since the field is bigger than our entire planet, overall, Earth's magnetic power is formidable.

Does Earth's **magnetic field** ever **change**?

Yes, the magnetic field is constantly changing, though very slowly. The magnetic poles actually drift several kilometers each year, often in seemingly random directions. Over thousands of years, the strength of the magnetic field can go up and down significantly. Even more amazing, Earth's magnetic field can reverse directions—the north magnetic pole becomes the south magnetic pole, and vice versa. According to scientific measurements, our planet's magnetic field last had a polarity reversal about 800,000 years ago.

What **will happen** when Earth's **magnetic field flips** upside down?

Probably not much will happen to our daily lives when Earth's magnetic field undergoes a polarity reversal. Measurements over the years show that there has been about a six percent reduction in the strength of Earth's magnetic field in the past century, so some scientists think that a polarity reversal on Earth will likely happen sooner rather than later. Some non-scientific hypotheses have been put forth, suggesting that there will be an environmental catastrophe as a result. There is no scientific reason to believe, however, that such disasters will occur.

Do any **other objects** in the solar system have **magnetic fields** that **flip upside down**?

Yes, all planets and stars with magnetospheres are thought to undergo magnetic polarity reversals. The Sun, for example, undergoes a magnetic field polarity reversal every 11 years. Astronomers can see and study this effect in other astronomical bodies, and from them, learn more about the changes in Earth's own magnetic field.

VAN ALLEN BELTS

What are the **Van Allen belts**?

The Van Allen belts are two rings of electrically charged particles that encircle our planet. The belts are shaped like fat doughnuts, widest above Earth's equator and curving downward toward Earth's surface near the polar regions. These charged particles usually come toward Earth from outer space—often from the Sun—and are trapped within these two regions of Earth's magnetosphere.

Since the particles are charged, they spiral around and along the magnetosphere's magnetic field lines. The lines lead away from Earth's equator, and the particles shuffle back and forth between the two magnetic poles. The closer belt is about 2,000 miles (3,000 kilometers) from Earth's surface, and the farther belt is about 10,000 miles (15,000 kilometers) away.

How were the **Van Allen belts discovered**?

In 1958 the United States launched its first satellite, *Explorer 1,* into orbit. Among the scientific instruments aboard *Explorer 1* was a radiation detector designed by James Van Allen (1914–2006), a professor of physics at the University of Iowa. It was this detector that first discovered the two belt-shaped regions of the magnetosphere filled with highly charged particles. These regions were subsequently named the Van Allen belts.

Do **other objects** in the solar system have **Van Allen belts**?

Yes. All the gas giant planets are thought to have such belts, and in Jupiter's magnetic field such belts have been observationally confirmed.

NEUTRINOS

What is a **neutrino**?

A neutrino is a tiny subatomic particle that is far smaller than an atomic nucleus; it has no electrical charge and a tiny mass. (Electrons are many thousands of times more massive than neutrinos, and protons and neutrons are many millions of times more massive.) Neutrinos are so tiny and ghostly that

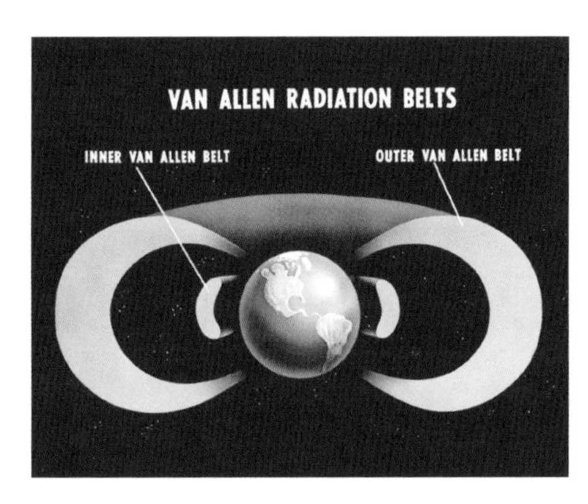

An illustration depicting the two layers of Van Allen Belts around the Earth. (*NASA*)

Am I getting hit by neutrinos right now?

You—and every square inch of Earth's surface—are being continuously bombarded by neutrinos from space. Billions of neutrinos slice through your body every second.

Fortunately, neutrinos are so unlikely to interact with any matter—including the atoms and molecules in the human body—that the billions upon billions of neutrinos that hit you every second have no discernible effect at all. In fact, the odds that any neutrino striking Earth will interact with any atom in our planet at all is about one in a billion. Even when it does happen, the result is merely a tiny flash of harmless light.

they almost always pass through any substance in the universe without any interference or reaction.

How was the **existence** of **neutrinos proven**?

The existence of neutrinos was first suggested in 1930 by the Austrian physicist Wolfgang Pauli (1900–1958). He noticed that in a type of radioactive process called beta decay, the range of the total energy given off in observations was greater than theoretical predictions. He reasoned that there must be another type of particle present to account for, and carry away, some of this energy. Since the amounts of energy were so tiny, the hypothetical particle must be very tiny as well and have no electric charge. A few years later, the Italian physicist Enrico Fermi (1901–1954) coined the name "neutrino" for this enigmatic particle. The existence of neutrinos was not experimentally confirmed, however, until 1956, when American physicists Clyde L. Cowan, Jr. (1919–1974) and Frederick Reines (1918–1998) detected neutrinos at a special nuclear facility in Savannah River, South Carolina.

If **neutrinos** are so **elusive**, how do **scientists observe** them striking Earth?

It is possible to detect neutrinos from space by their very rare interactions with matter here on Earth, but not with conventional telescopes. The first effective neutrino detector was set up in 1967 deep underground in the Homestake Gold Mine near Lead, South Dakota. There, the American scientists Ray Davis, Jr. (1914–) and John Bahcall (1934–2005) set up a tank filled with 100,000 gallons of nearly pure perchlorate (used as dry-cleaning fluid), and monitored the liquid for very rare neutrino interaction events. Other experiments have since used other substances, such as pure water, for neutrino detections.

Where are the **neutrinos coming** from?

The vast majority of neutrinos striking our planet come from the Sun. The nuclear reactions at the core of the Sun create huge numbers of neutrinos; and unlike the

What was the "solar neutrino problem?"

From the very beginning of neutrino astronomy research, there was a discrepancy between the theory of nuclear fusion and the number of neutrinos detected from the Sun. Neutrino telescopes on Earth detected only about half as many neutrinos as they should have. This strange result was checked again and again and repeatedly confirmed. This became known as the solar neutrino problem. Was the Sun generating less energy at its core than expected? Was nuclear fusion theory wrong?

The problem was finally solved nearly four decades after it was first discovered. Neutrinos, as it turns out, can actually change their characteristics when they strike Earth's atmosphere. That meant that there were the right number of neutrinos leaving the Sun, but so many of them changed "flavor" upon reaching Earth that they escaped detection by the neutrino telescopes deep underground. This discovery was a major breakthrough in fundamental physics. It confirmed very important properties about neutrinos that have major implications on the basic nature of matter in the universe.

light that is produced, which takes thousands of years to flow their way out of the Sun's interior, the neutrinos come out of the Sun in less than three seconds, reaching Earth in just eight minutes.

Have **neutrinos** ever been shown to have hit Earth **from somewhere other than the Sun**?

In 1987, the first supernova visible to the unaided eye to occur in centuries appeared in the southern sky. At almost exactly that same moment, neutrino detectors around the world recorded a total of 19 more neutrino reactions than usual. This worldwide detection does not sound like much, but it was hugely significant because it was the first time neutrinos were confirmed to have reached Earth from a specific celestial object other than the Sun.

COSMIC RAYS

What are **cosmic rays**?

Cosmic rays are invisible, high-energy particles that constantly bombard Earth from all directions. Most cosmic rays are protons moving at extremely high speeds, but they can be atomic nuclei of any known element. They enter Earth's atmosphere at velocities of 90 percent the speed of light or more.

Am I getting hit by cosmic rays?

Everyone is being struck by cosmic rays all the time—probably about several each second. Ordinarily, the number of cosmic rays that strike you have no negative effect on your health. Even though the energy of these particles is very high, the number of them striking you is relatively low. If you went beyond Earth's magnetosphere, though, your health might be at risk. On Earth's surface, the magnetosphere acts as a shield against cosmic rays by redirecting them toward Earth's magnetic poles. Thousands of miles up, however, the cosmic ray flux on your body would be much higher, and thus cause potentially more damage to your body's cells and systems.

Who **first discovered cosmic rays**?

The Austrian-American astronomer Victor Franz Hess (1883–1964) became interested in a mysterious radiation that scientists had found in the ground and in Earth's atmosphere. This radiation could change the electric charge on an electroscope—a device used to detect electromagnetic activity—even when placed in a sealed container. Hess thought that the radiation was coming from underground and that at high altitudes it would no longer be detectable. To test this idea, in 1912 Hess took a series of high-altitude, hot-air balloon flights with an electroscope aboard. He made ten trips at night, and one during a solar eclipse, just to be sure the Sun was not the source of the radiation. To his surprise, Hess found that the higher he went, the stronger the radiation became. This discovery led Hess to conclude that this radiation was coming from outer space. For his work on understanding cosmic rays, Hess received the Nobel Prize in physics in 1936.

How were **cosmic rays** shown to be **charged particles**?

In 1925 American physicist Robert A. Millikan (1868–1953) lowered an electroscope deep into a lake and detected the same kind of powerful radiation that Victor Franz Hess had found in his balloon experiments. He was the first to call this radiation cosmic rays, but he did not know what they were made of. In 1932, the American physicist Arthur Holly Compton (1892–1962) measured cosmic-ray radiation at many points on Earth's surface and found that it was more intense at higher latitudes (toward the North and South Poles) than at lower latitudes (toward the equator). He concluded that Earth's magnetic field was affecting the cosmic rays, deflecting them away from the equator and toward Earth's magnetic field. Since electromagnetism was now shown to affect the rays, it was clear that cosmic rays had to be electrically charged particles.

Where do **cosmic rays** come **from**?

A continuous stream of electrically charged particles flows from the Sun; this flow is called the solar wind. It makes sense that some fraction of cosmic rays originate from the Sun, but the Sun alone cannot account for the total flux of cosmic rays onto Earth's surface. The source for the rest of these cosmic rays remains mysterious. Distant supernova explosions could account for some of them; another possibility is that many cosmic rays are charged particles that have been accelerated to enormous speeds by interstellar magnetic fields.

METEORS, METEORITES, ASTEROIDS, AND COMETS

What is a **meteorite**?

A meteorite is a large particle from outer space that lands on Earth. They range in size from a grain of sand on up. About 30,000 meteorites have been recovered in recorded history; about 600 of them are made primarily of metal, and the rest are made primarily of rock.

What is a **meteor**?

A meteor is a particle from outer space that enters Earth's atmosphere, but does not land on Earth. Instead, the particle burns up in the atmosphere, leaving a short-lived, glowing trail that traces part of its path through the sky. Like meteorites, meteors can range from the size of a grain of sand on up; most of the time, though, a meteor larger than about the size of a baseball will reach Earth, in which case we call it a meteorite.

Barringer crater in Arizona is clear physical evidence of the power of a meteor impact. (*NASA*)

Where do **meteors and meteorites** come **from**?

Most meteors, especially those that fall during meteor showers, are the tiny remnants of comets left in Earth's orbital path over many, many years. Most meteorites, which are generally larger than meteors, are pieces of asteroids and comets that somehow came apart from their parent bodies—perhaps from a collision with another body—and orbited in the solar system until they collided with Earth.

What is a **meteor shower**?

Meteors are often called "shooting stars" because they are bright for a moment and move quickly across the sky. Usually, a shooting star appears in the sky about once an hour or so. Sometimes, though, a large number of meteors appear in the sky over the course of several nights. These meteors will appear to come from the same part of the sky, and dozens or hundreds (sometimes even thousands) of meteors can be seen every hour. We call such dazzling displays meteor showers. The strongest meteor showers are sometimes called meteor storms.

What is a **bolide**?

Bolides are meteors large enough to create an easily-seen fireball as they enter Earth's atmosphere. They can even be seen during the daytime and often appear greenish, though other hues have also been observed.

Are **falling meteors** and **meteorites dangerous**?

Typical meteors and meteorites pose no danger of any kind to people. Meteors burn up before they reach Earth, so they do not hit anything on the surface; meteorites are so rare that the chances of their hitting anything important are almost zero.

Still, occasional incidents are known to happen. A falling meteorite killed a dog in Egypt in 1911; another struck the arm of—and rudely awakened—a sleeping woman in Alabama in 1954; and in 1992 a meteorite put a hole through a Chevy Malibu automobile. Once in a very rare while—every 100,000 years or so—a meteor or meteorite about 300 feet (100 meters) across will collide with Earth. Once in a very, very rare while—every 100 million years or so—a meteorite 3,000 feet (1,000) meters across will do so, and that is a cataclysmic event.

What is the **largest known meteorite** to **strike Earth** in the past **100,000 years**?

About 50,000 years ago, a metallic meteorite about 100 feet (30 meters) across crashed into the Mogollon Rim area in modern-day Arizona. It disintegrated on impact, creating a hole in the desert nearly a mile across and nearly 60 stories deep. Meteor Crater (or the Barringer Meteor Crater, as it is more commonly known today) is a remarkable and lasting example of the amount of kinetic energy carried by celestial objects. Just the lip of the crater rises 15 stories up above the desert floor. For a long time, scientists puzzled over the origin of this crater. It might have been vol-

What is the largest known meteor to disintegrate in Earth's atmosphere in recent times?

On the night of June 30, 1908, villagers near the Tunguska River in Siberia witnessed a fireball streaking through the sky, a burst of light, a thunderous sound, and an enormous blast. A thousand miles away, in Irkutsk, Russia, a seismograph recorded what appeared to be a distant earthquake. This area was so remote, however, that a scientific expedition to the site did not happen until 1927. Incredibly, they found more than 1,000 square miles of burned and flattened forest.

Modern scientific computations have shown that this incredible explosion was probably caused by a small, rocky asteroid or comet about 100 feet (30 meters) across. Computer simulations show that it most likely came into Earth's atmosphere at a shallow angle and exploded in mid-air above the forest. The explosion packed a punch easily greater than 1,000 Hiroshima atomic bombs.

canic in origin, they thought. But geological evidence, such as shallow metallic remnants in a huge radius miles around the crater, confirmed it was a meteorite strike.

What is the **largest known meteorite** to **strike Earth** in the past **100 million years**?

About 65 million years ago, a meteorite about 6 miles (10 kilometers) across crashed into our planet near what is now southern Mexico. The remnant of this collision is an underwater crater more than 100 miles (161 kilometers) across. This asteroid, or comet, carried ten million times more kinetic energy than either the Tunguska or Meteor Crater impactors. The heat from the explosion probably set the air itself on fire for miles around. It threw so much of Earth's crust into the atmosphere that it blocked most of the Sun's light for months. As it fell back through the atmosphere, this debris grew very hot as it landed; it probably set almost every tree, bush, and blade of grass it touched on fire. The ecological catastrophe caused by this titanic meteorite strike was most likely the evolutionary blow that finished off the dinosaurs.

What is an **asteroid**?

Asteroids are relatively small (compared to moons and planets), rocky objects in our solar system. They range in size from a few feet across to behemoths like Ceries, which is 580 miles (933 kilometers) in diameter. The majority of asteroids can be found in the Asteroid Belt, a band of rocks orbiting between Mars and Jupiter. The origin of asteroids remains the subject of scientific study. Astronomers today think that most asteroids are planetesimals that never quite combined with other bodies to form planets. Some asteroids, on the other hand, may be the shattered remains of planets or protoplanets that suffered huge collisions and broke into pieces. Aster-

Are all asteroids located in the asteroid belt?

No. There are many asteroids in other regions of the solar system. Chiron, for example, which was discovered in 1977, orbits between Saturn and Uranus. Another example is the Trojan asteroids that follow the orbit of Jupiter near Lagrange points—one group preceding the planet, the other following it—and can thus orbit safely without crashing into Jupiter itself.

oids range in composition from "rubble piles" that are loose collections of rock held together by gravity (e.g., the Mathilde asteroid) to solid rock (Eros) to asteroids with high metallic content (Kleopatra). Some asteroids are so large that they exert gravitation and have tiny moons orbiting them.

What are **near-Earth objects** and are they **dangerous**?

There are hundreds, if not thousands, of NEOs—near-Earth objects, which are asteroids with orbits that cross Earth's orbit. An NEO could indeed strike our planet, possibly unleashing cosmic destruction. As of early 2009, astronomers have tracked nearly 6,000 near-Earth objects. Of these, 765 have diameters of more than a mile across, and over 1,000 are considered Potentially Hazardous Asteroids (PHAs), meaning that their orbits take them uncomfortably close to Earth.

Has an **asteroid** ever **hit Earth**?

Yes. In fact, the extinction of the dinosaurs is believed to have resulted from an asteroid striking the Earth near the shoreline of what is now the Yucatan Peninsula in Mexico. There is also speculation that, 13,000 years ago, a comet or large asteroid smashed into the planet and wiped out the Native American Clovis civilization, as well as mastodons, mammoths, and other large animal species that once roamed North America. This sort of impact is, of course, extremely rare. Today, astronomers estimate that small asteroids approach Earth and burn up in the atmosphere at a rate of about two or three a year. In October 2008, scientists were excited to have accurately predicted that a small asteroid (2008 TC3), measuring about 15 feet (4.5 meters) across, would enter the atmosphere. It was the first time that NASA's Near-Earth Object Program had successfully anticipated such an event.

What is a **comet**?

Comets are basically "snowy dirtballs" or "dirty snowballs"—clumpy collections of rocky material, dust, and frozen water, methane, and ammonia that move through the solar system in long, highly elliptical orbits around the Sun. When they are far away from the Sun, comets are simple, solid bodies; but when they get closer to the Sun, they warm up, causing the ice in the comets' outer surface to vaporize. This creates a cloudy "coma" that forms around the solid part of the comet, called the “nucleus.” The loosened comet vapor forms long “tails” that can grow to millions of miles in length.

Comet C/2002 V1 (also known as NEAT) as seen in a Kitt Peak Observatory photograph. (*NASA*)

What do some people fear will occur in **December 2012**?

There is currently a growing number of doomsday believers who believe the world will end in December 2012. They explain that prophecies stemming from sources as diverse as the Mayan calendar and Nostradamus predict that human civilization will either be destroyed or dramatically changed at that time because a comet will strike the planet. Biblical numerologist Harold Camping has made a similar prediction, except that he places the date at October 21, 2011. Astronomers tracking comets, however, have found no evidence that any of the comets in our solar system are heading our way any time soon.

From **where** do **comets originate**?

Most of the comets that orbit the Sun originate in the Kuiper Belt or the Oort Cloud, two major zones in our solar system beyond the orbit of Neptune. “Short-period comets” usually originate in the Kuiper Belt. Some comets and comet-like objects, however, have even smaller orbits; they may have once come from the Kuiper Belt and Oort Cloud, but have had their orbital paths altered by gravitational interactions with Jupiter and the other planets.

What would happen if a large **comet or asteroid impact** occurred?

If a large extraterrestrial body, such as a comet or asteroid, with a diameter of six miles (10 kilometers) or more struck anywhere on our planet, land or sea, the impact would be felt worldwide. The initial impact would vaporize parts of the Earth’s crust, as well as the oceans, and send shockwaves out to every corner of the globe. Debris would be shot out into the upper atmosphere, and the heat generated from the collision with the atmosphere and crust would turn the sky red and set the world’s forests aflame. After the initial collision, debris would rain down everywhere, destroying even those buildings, wildlife, and people who were far from the impact. Virtually all living things inhabiting the surface would be killed, and even sea creatures would die as the temperatures of the oceans reached the boiling point. Long after the impact, the atmosphere would be coated in a layer of ash that would circle the Earth for months or even years, plunging the planet into a “nuclear winter” after the initial fireball. About the only things that might survive—such as is believed to have happened 65 million years ago after the last such event—would be creatures that could burrow underground or hide deep inside caves.

Can **smaller asteroids and comets** affect the weather?

Certainly. It doesn't take a huge impact for an extraterrestrial body to influence the planet's delicate weather balance. Smaller impacts could still throw up a considerable amount of dust, comparable to a large volcano exploding. It doesn't take a very large asteroid to do the trick, either. Scientists now estimate that the object that hit Tunguska, Russia, may have been as small as 65 feet (20 meters) in diameter.

But an asteroid or meteorite doesn't even have to hit the Earth to affect our weather. In 2005, scientists announced that asteroids burning up in the atmosphere leave behind trails of micron-sized particles that could increase cloud formation, increase precipitation, and cool temperatures.

HUMANITY AND THE WEATHER

HUMANITY'S IMPACT

Has the **weather** ever been responsible for **killing a U.S. president**?

When U.S. President William Henry Harrison (1773–1841) gave his inaugural address on March 4, 1841, the weather was miserable. Despite snow and bitter cold, President Harrison insisted on remaining outside and giving one of the longest inaugural addresses ever (1 hour and 45 minutes). At 68 years old, the former general and war hero was not in the most robust health, and many blamed the weather conditions as the cause of President Harrison's subsequent pneumonia. One month after becoming president, he died, thus earning him the distinction of the U.S. president serving the shortest term in office.

What are the **consequences** in the United States when there is an **unusually cool summer**?

Cooler than normal summers can have both benefits and drawbacks. Drawbacks include poorer crop harvests and, in states where summer tourism is important, reduced income for businesses and fewer tax dollars for local government. People tend to run their air conditioners less, as well, which is good for the environment but hard on utility companies trying to maintain a profit. Meanwhile, consumers save on utility expenses. As one might guess, incidents of heat stroke and dehydration are reduced and hospitals see fewer patients stricken by these problems.

Is Earth's **atmosphere changing**?

Earth's atmosphere is continually and gradually changing. Over cycles that usually last many thousands of years, the concentrations of different gases—including oxygen, carbon dioxide, and others—go up and down, as does the concentrations of tiny dust particulates such as carbon soot.

In the past hundred years or so, human population growth and industrial activity has caused a much sharper change in the concentrations of some gases and particulates on a much shorter timescale than at any time in the past 200,000 years. The most dramatic effect has been a huge increase in the amount of carbon dioxide in the atmosphere. This increase has created a substantial greenhouse effect, according to some scientists, which could be increasing the average temperature on Earth at a much faster rate than typical ecological and geological timescales.

What is **weather modification**?

Since the advent of agriculture, when food supplies waxed or waned depending on the weather, human civilizations have wanted to control the weather. Ancient civilizations typically did this by praying to gods or to one God to bring rain and prosperity. In the twentieth century, however, science finally seemed to offer hope that our understanding of chemistry and meteorology could be used to affect the weather. There was considerable hope that cloud seeding, discovered in 1946, could make droughts a thing of the past. The potential of this technique never met expectations, though it still can be used under the right conditions; seeding techniques are also employed to reduce the size of hail stones so they will cause less damage. The Weather Modification Association, which was established in 1950, is still active today and supports research in this area.

What is **inadvertent weather modification**?

This, of course, means accidentally changing the weather, usually for the worse. Humanity has been very good at that, especially as a result of activities that lead to deforestation and pollution. It is now common knowledge that chemicals from cars and industry are destroying the ozone, and particulates from emissions also have an effect on precipitation. Cutting down forests in favor of croplands, urban construction, golf courses, and so on causes more sunlight to be reflected back into the atmosphere but can also cause urban heat islands. Animal agriculture can increase methane gas in the atmosphere, and irrigation of plants alters water distribution. All of these activities and more have noticeably changed natural weather patterns and the climate without humans intending to do so.

Who first theorized that **human activities** were causing **climate change**?

The Swedish scientist Svante August Arrhenius (1859–1927), who was one of the founders of the discipline of physical chemistry, is sometimes credited as the first person to discuss in detail the effects of carbon dioxide (CO_2) on Earth's climate. Arrhenius came upon his theory while studying past ice ages, and in 1896 published a paper in which he proposed that ice ages occur when CO_2 levels go down. By his estimates, doubling CO_2 amounts would increase average world temperatures by about 2.5°F (5°C) and halving the levels of this gas would have the opposite effect.

Recent modeling studies come quite close to Arrhenius's original estimate. Unlike today's environmentalists and climatologists, however, the Swedish scientist believed that global warming would be a good thing for two reasons: it would help prevent another ice age and it would aid in crop production to feed a hungry world.

Farmers in Brazil use fires to clear away forest, which can easily be seen in this satellite photo. (*NASA*)

Why are **trees and other vegetation** important to the weather and climate?

Cutting down trees, especially on the massive scale that is occurring today, has a definite effect on the weather in several ways: 1) trees and other plant life absorb carbon dioxide and other pollutants that cause global warming; 2) plants also absorb sunlight and areas covered by forest thus reflecting less sunlight back into the atmosphere; 3) trees near homes and commercial buildings help reduce electricity use because they keep buildings cooler in the summer and serve as a barrier against cold winds in the winter.

What is the current extent of **deforestation**?

Worldwide, we are losing enough forestland annually to cover an area the size of the state of Panama. To put that figure into numbers, that is a net loss of 7.3 million hectares (18 million acres) every year. Actually, about 13 million hectares (32 million acres) are being chopped down yearly, but there are restoration projects going on that help replant forests. This rate of loss (which is an average taken from the years 2000 through 2005) is somewhat better than the 8.9 million hectare (22 million acre) loss that was experienced during the previous decade. Although replanting is great, new forest growth does not make for as healthy a habitat for wildlife as old forest growth.

GENERAL POLLUTION FACTS

How is **pollution** related to **weather**?

Pollution—both natural and man-made—has numerous, often complex effects on the weather. Air pollutants, for instance, can cause acid rain, and pollution that destroys the ozone layer can lead to health risks to people and even wipe out species. Many scientists believe that human-generated pollution is causing climate change, which affects weather patterns on a global scale. While some pollutants, such as gases from volcanic eruptions, can have harmful consequences, many meteorologists, environmentalists, and climatologists fear that human activity is having a much bigger, negative impact on the weather and our health than anything currently occurring from natural sources.

What is **long range transport**?

Long range transport refers to the fact that winds—especially high-altitude winds—can carry pollutants over incredibly long distances. People once believed that pollutants from sources such as smoke stacks might travel a few miles before settling onto the ground and water supplies. Now it is known that these particulates and toxic gases can make their way into the upper atmosphere. Scientists first started to become aware of this issue in the mid-twentieth century, when nuclear bomb tests resulted in radioactive clouds that would circumnavigate the planet. Acid-rain-causing chemicals can easily cross the entire United States, and, likewise, pesticides and herbicides travel long distances. Natural air pollutants, such as volcanic ash and organic material like fungi, spores, and pollen, can similarly range over long distances.

What **major energy source** is considered to be the **cleanest**?

Natural gas is regarded as the cleanest-burning fossil fuel, producing less pollution than oil or coal.

What is an **urban heat island**?

Because urban areas are generally devoid of significant vegetation, the concrete and other construction materials for buildings and roads prevent the heat from the Sun from being absorbed. Instead, surfaces become hotter and drier. Cities and towns become warmer than surrounding rural areas to the point where, on a warm summer day, surfaces such as sidewalks and roofs can be as much as 50° to 90°F (27 to 50°C) hotter than the surrounding air. The warming effects are especially pronounced during the day, but the temperatures are also affected at night.

According to the U.S. Environmental Protection Agency, a city of about one million people can cause the surrounding atmosphere to heat up as much as 22°F (12°C) more than it would be under similar weather conditions in a rural area. Annually, the effect would be that the overall temperature for a city that size would be 1.8 to 5.4°F (1 to 3°C) higher than in the surrounding areas. There are several problems that result from heat islands: people tend to use their air conditioners and other utilities more, thus increasing energy consumption; this leads to more pollution, including greenhouse gases; these pollutants also affect human health; finally, rainwater that has fallen onto heated surfaces on pavement and roofs flows into sewers, and then into the environment, where the heated water affects wildlife.

What is an **urban ice slab**?

Just as urban areas can cause problems related to heat, they also can make winters more hazardous. When ice forms on skyscrapers and other tall buildings, a dangerous situation may arise upon thawing. Large sheets of ice have been known to break

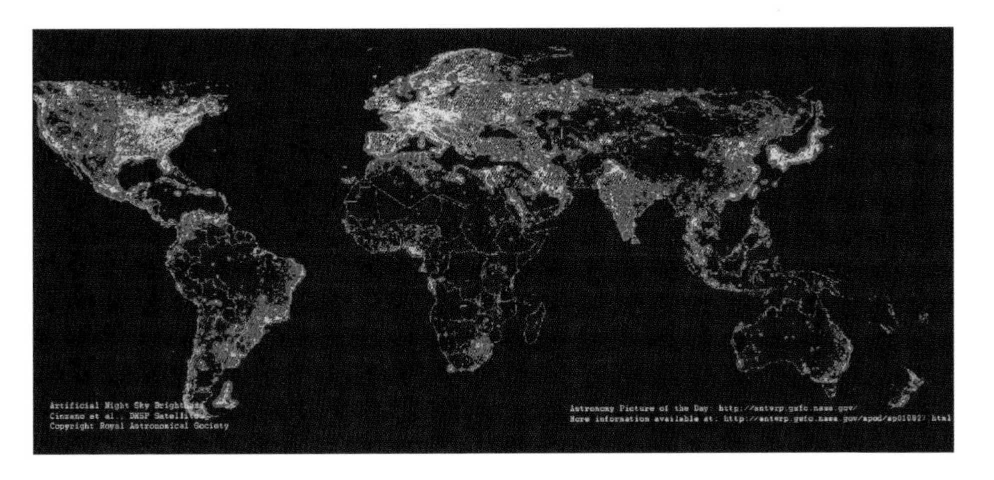

It's easy to see the extent of light pollution in this image of nighttime lights across the globe. (*NASA*)

off of structures and plummet to the streets below. One notable case of this happening occurred in Chicago, Illinois, in April 1995, when the danger of urban ice slabs dropping onto Michigan Avenue compelled authorities to close the street for hours.

What is **light pollution**?

While not harmful to people or other living things, light pollution is an annoyance to astronomers. Light from cities makes it difficult to see stars and other heavenly bodies at night, which is why observatories are stationed on top of mountains and hills outside of urban areas. It is also why space observatories such as the Hubble Telescope are so important for astronomers.

What is **odor pollution**?

Odor pollution is simply the unpleasant stench of such things as garbage, sewage, chemicals, rotting organic matter, and hazardous waste. Human beings can detect foul odors at concentrations as low as one part per trillion, depending on what the chemical is. This is important because people can actually detect air pollutants (except for odorless ones such as carbon monoxide) at levels much lower than can be detected by instruments. This is especially true of the smellier pollutants, such as hydrogen sulfide (H_2S), which produces that familiar rotten egg smell, even when diluted in water. Hydrogen sulfide pollution can cause everything from eye and throat irritation to asthma attacks and even death.

AIR POLLUTION

What is **air pollution**?

Air pollution is caused by many sources. There are natural pollutants that have been around as long as the Earth, such as dust, smoke, volcanic ash, and pollens.

Humans have added to air pollution with chemicals and particulates due to combustion and industrial activity (anthropogenic sources).

What is **smog**?

The word "smog" is a combination of two other words: smoke and fog. Harold Des Voeux, a British physician who was concerned about air quality, is credited with coming up with coining the term in 1911. However, what we refer to as smog has nothing to do with either fog or smoke, usually. Smog is simply another name for air pollution. Scientists refer to it, more precisely, as photochemical smog because it is the result of chemical reactions in the presence of sunlight. The brownish haze associated with smog is the result of nitrogen dioxide in the air, but smog also includes a soup of nitrogen oxides, hydrocarbons, aldehydes, ozone, peroxyethanoyl nitrates (PANs), and suspended particulate matter. Yuck.

What is a **photochemical grid model**?

A photochemical grid model is a computer model used by meteorologists and environmental scientists to simulate what might happen during air pollution episodes under various weather conditions. A grid system is employed in which the area of study—for instance, a city—is cut up into thousands of cells, each usually a couple miles or kilometers wide and long; the cells also have a third dimension (height), varying in depth according to the altitude that scientists wish to study. These models simulate vertical and horizontal air movements, increases in various gases and particles from sources ranging from buildings and cars to plants and animals, and chemical reactions occurring in the atmosphere; they are very useful in predicting effects on ozone levels. A photochemical grid model is thus not the same as a meteorological model, but it does use this meteorological tool to study how pollution increases, dissipates, and affects certain areas. Photochemical grid models can be used to simulate how making different decisions affecting pollutant outputs would affect air quality. For example, if city officials decided to limit commuter traffic into the downtown area by ten percent, they could simulate how levels of carbon monoxide would be reduced.

How much **carbon monoxide** is produced in the United States by **cars**?

As of 2002, automobile traffic in the United States was producing 346 tons (314 metric tons) of carbon monoxide per day.

Is **carbon dioxide pollution** a problem?

While carbon dioxide occurs naturally in the environment, and, indeed, is essential to plant life, too much of even a good thing can be bad. Increased carbon dioxide levels, of course, are now infamous for leading to global warming (see the chapter on climate change for more on this), but near ground level this gas can also be poisonous to plants and animals. This became readily apparent during a 1990 incident in which carbon dioxide emissions from volcanic faults in California's Inyo Nation-

What U.S. city is particularly noted for its problems with smog?

Los Angeles has a rough time with smog. The brown haze that often lingers over the city is a result of several factors. Of course, the urban area is filled with cars and other sources of pollutants, but the natural environment conspires to make matters worse. First of all, very little rain falls in L.A., which might attract tourists and new residents but does nothing to wash away pollutants; and secondly, the city is in a basin surrounded by mountains. Ocean breezes blowing in from the west keep air pollution from escaping in that direction, but then the smog finds another barrier in the mountains that lie east, north, and south of the city. Even before Spanish and other European settlers came to the area, the Native American Chumash tribe called what is now Los Angeles the "valley of smoke," because haze from brush fires and dust would be trapped there for long periods.

al Forest killed trees and caused tourists to become dizzy and lightheaded. The tourists became ill while inside cabins, where the levels of carbon dioxide had increased to 25 percent of the air content.

What is a **brown cloud**?

The brown haze suspended over cities such as Los Angeles, Mexico City, and Cairo, Egypt, is sometimes referred to as a "brown cloud."

What are **pollution hot spots**?

Levels of pollution are not consistent throughout polluted areas such as industrialized cities. For example, car emissions along freeways raise air pollutants significantly in adjacent areas, and air quality inside tunnels and parking garages is considerably worse because of limited ventilation. Other potential hot spots include areas, whether urban or rural, that lie downwind of factories and power plants.

What are the **most polluted cities** in the world in terms of **air quality**?

The most severe air pollution can be found in the following cities: Cairo, Egypt; Delhi, India; Calcutta, India; Tianjin, China; Chongqing, China; Kanpur, India; Lucknow, India; Jakarta, Indonesia; and Shenyang, China.

What are some **disturbing statistics** regarding **air pollution in Egypt**?

The World Health Organization estimated that breathing the air in Cairo, Egypt, is the toxic equivalent of smoking 20 cigarettes every day. Damage to the Egyptian economy as a result of pollution is also troubling, with the World Bank noting in

Cairo, Egypt, is one of the smoggiest cities on the planet.

2002 that five percent of the country's gross domestic product (or about $2.42 billion U.S. dollars) are lost annually due to damages attributed to pollution.

How much **air pollution** does the **United States produce**?

According to the Environmental Protection Agency, there is some good news in that air pollution has been slowly declining in the United States. For example, carbon monoxide emissions have gone down from 178 million tons (161.5 metric tons) in 1980 to 81 million tons (73.5 million metric tons) in 2007. Volatile organic compounds (VOCs) and sulfur dioxide have both been cut in half (from 30 to 15 million tons [27.2 to 13.6 metric tons] for VOCs and 26 down to 13 million tons [23.6 to 11.8 metric tons] for SO_2) over the same time period, and nitrogen oxides have been lowered from 27 to 17 million tons (24.5 to 15.4 metric tons). Overall, from 1980 to 2007, air pollutant rates have dropped from 267 million tons (242 metric tons) produced annually to 129 million tons (117 metric tons). This is still a lot of pollution, without question, but it is significant progress, especially considering that the U.S. population has risen from about 226 million people in 1980 to 300 million in 2007.

Why is **air pollution in China** such a huge problem?

In recent years, China has been experiencing an unprecedented economic and industrial boom. Cities are growing, manufacturing (with the exception of the onset of a worldwide recession in 2008) has been flourishing, and living standards are rising. This has been great news for many Chinese citizens, but it has come with a cost: a huge environmental problem. Although China has many environmental regulations on the books, it has had tremendous problems enforcing them. The result is that the country is probably the most polluted on the planet, with only India as a rival in this

What was the Donora Smog Disaster of 1948?

One of the most shameful chapters in environmental history occurred on October 30 and 31, 1948, in the industrial town of Donora in Washington County, Pennsylvania. The 14,000 residents were mostly supported by steel mills, which found the location by the Monongahela River just 30 miles (48 kilometers) from Pittsburgh to be ideal. The mills provided good-paying jobs, but the blast furnaces they used produced massive quantities of soot, sulfur dioxide, and other pollutants. In addition, the town was home to a zinc smelting plant and a sulfuric acid plant. As if that weren't enough, the natural climate of the town made foggy weather a regular occurrence. The industry combined with the weather during a horrible Halloween season in 1948, creating such thick, brown air that it was nearly impossible to see where one was going. Five hundred of the residents fell ill from various respiratory problems and 22 people died (17 from complications from asthma or heart disease, and two from tuberculosis aggravated by the filthy air).

infamous area. The air quality in about two thirds of the cities is considered unhealthy for people, acid rain is damaging crops, and life expectancy rates are dropping because of pollution-related health issues. The Chinese government is not unaware of this shameful problem, and the country made news in 2008 when it hosted the Olympic Games. Heavy restrictions were placed on auto traffic and factories so that the air would be clear enough that the athletes would not be sickened by the smog.

What are the **sources** of **air pollution**?

Air pollution can be in the form of either a gas or an aerosol, and it can either be anthropogenic (man-made) or natural. Man-made sources of pollution include factories, cars, motorcycles, ships, incinerators, wood and coal burning, oil refining, chemicals, consumer product emissions like aerosol sprays and fumes from paint, methane from garbage in landfills, and pollution from nuclear and biological weapons production and testing.

How much **pollution** is generated by **airplanes**?

At any given moment in time, there are about 5,000 non-military aircraft plying the skies over the United States. It takes a tremendous amount of fuel to keep all these transports aloft, and the burned fuel results in exhaust similar to that coming from car engines: nitrogen oxides, sulfur dioxide, carbon monoxide, and soot. In addition, water vapors are released, which form ice crystals at high elevations commonly called contrails. Meteorologists believe that these contrails increase the formation of cirrus clouds, which may contribute to global warming.

Large tracts of land in the Amazon rain forest are regularly deforested to make way for farms, but, ironically, the soil is not very fertile.

Does **air pollution** reach as far as the **North Pole**?

Yes. Winds can carry air pollutants far beyond the Arctic Circle, resulting in a condition called "Arctic haze." The pollution tends to be worse during winter and spring, when prevailing winds from northern Europe to Siberia blow emissions northward from industrial areas. Recent shifts from coal burning to natural gas (primarily from Russia) has created cleaner air conditions, fortunately.

What are some **natural sources** of **air pollution**?

Natural sources of pollution may include dust, methane from human and animal waste or flatus, radon gas, smoke from wildfires, and volcanic activity.

Are there any circumstances where **air pollution** could be **beneficial**?

Yes. In fact, dust storms and volcanic ash can be a means for nature to distribute soils across the surface of the planet to regions that might not otherwise get certain nutrients. For instance, the Amazon rain forest may be famous for its verdant plant life, but in actuality the soil beneath the dense canopies is of very poor quality. Dust blown across the ocean from Africa to South America may help fertilize the rain forests. Not only that, but such storms bring nutrients to plankton in the ocean, which form the basis for aquatic food chains.

Can **weather** actually make **pollution worse**?

Yes, and it can alleviate it, as well. Rain, for instance, can wash away haze over cities, and wind can blow it away. Stagnant air masses, humid air, or temperature inver-

Did air pollution originate with the Industrial Age of the eighteenth century and beyond?

No. Since before recorded history, when humans first discovered fire, people have been generating pollution. As people learned to burn coal, they were also subjected to sulfur dioxide poisoning. Indeed, even Native Americans such as the Hopi used coal as a source of fire and heat.

sions, on the other hand, will allow pollution to build up. During the night, "nocturnal inversions" cause carbon monoxide to build up around freeways and other high-traffic areas. On the other hand, "mixed layer" conditions help to disperse pollutants when temperatures decrease over several thousand feet at a rate of about 4.5°F (2.5°C) for every thousand feet (about 300 meters) or so.

How much does **air pollution damage crops** in the United States?

It is estimated that the United States loses millions of dollars in crops every year because of air pollution. In the East, crop reductions cost about three billion dollars in losses annually, and farmland near such cities as Los Angeles and Chicago is far less productive than land far from such cities.

What was the **air quality** like in **nineteenth-century London**?

Words can barely describe how bad air pollution was during the late nineteenth century in London, England. Coal was burned in excess, and the soot and sulfur dioxide that resulted is blamed for a shockingly increased mortality rate among infants. Indeed, it is estimated that about 50 percent of the children born in London at the time failed to live past the age of two. So much sunlight was blocked by coal dust that people suffered from lack of vitamin D, with the result being a rise in rickets. And, of course, respiratory ailments were rampant.

What are **TSPs**?

Total suspended particulates (TSPs) are particles in the air ranging in size from 10 microns on down to less than a micron in diameter (a micron is one millionth of a meter). People can breathe in many TSPs, though larger particles can be weeded out by such things as nose hairs and mucous membranes. TSPs either result from primary sources (e.g., car exhaust, smokestacks) or are formed from secondary sources, such as ammonia and sulfur dioxide combining to form new pollutants (in this case, ammonium sulfate).

What are **VOCs**?

VOC stands for "volatile organic compound." They are organic chemicals that easily vaporize in the atmosphere.

What is **carbon monoxide**?

Carbon monoxide (CO) is an odorless, colorless, and tasteless gas that is lethal. Auto exhaust is one common source, but CO can result from the combustion of almost any material containing carbon. The molecules bond to hemoglobin in the blood, preventing the hemoglobin from transporting oxygen through the body as it normally does. Depriving organs and other tissues of oxygen can result in death within minutes. The early symptoms of carbon monoxide poisoning, however, include drowsiness, disorientation, and headaches.

In well-ventilated areas, carbon monoxide poisoning should not be a problem, but in closed-in areas, such as a garage, it is hazardous. This is why you should never leave your car running inside a garage. But carbon monoxide can also come from clogged chimneys, unvented space heaters, gas appliances, grills, and lawn mowers. Homes should be equipped with carbon monoxide monitors as a precaution.

While carbon monoxide poisoning is more likely inside a home or garage than outdoors, this pollutant can be a problem in large urban areas. In 1995, for example, there was a strong temperature inversion in the city of Chicago that caused carbon monoxide levels to be pushed toward the ground, rather than dissipating. The toxic gas then found its way into some homes.

What is **sulfur dioxide**?

Burning coal is a primary source of sulfur dioxide (SO_2) in the atmosphere, and so it was one of the first causes of air pollution, arising during the Industrial Age. Sulfur in bituminous and other forms of coal , when burned, bonds with oxygen to form this pollutant, which irritates eyes and the respiratory system. It is also a source of acid rain. Technologies have been developed that scrub sulfur dioxide emissions from smoke stacks, and have done much to improve air quality. While the United States has dramatically lowered these emissions, other developing nations, such as China and India, do not impose vigorous restrictions on sulfur dioxide from factories and power plants.

What are some other **significant air pollutants**?

Oxides of nitrogen, including nitrogen monoxide (NO), nitrogen dioxide (NO_2) and nitrous oxide (N_2O) are also generated by factory and car emissions. These gases are not directly harmful to people, but they contribute to ozone depletion.

What is the **Air Quality Index**?

The U.S. Environmental Protection Agency (EPA) developed the Air Quality Index as a measure to better advise citizens of pollutants in the atmosphere. To calculate the index, the EPA takes into account levels of carbon monoxide, ozone, nitrogen dioxide, and sulfur dioxide. Each pollutant is measured in parts per billion and compared to an acceptable standard over a certain period of time (24 hours for most pollutants, but ozone is calculated over an eight-hour period). This number is then multiplied by 100 to give the Air Quality Index. In other words, the formula would be (pollutant concentration)/(pollutant goal concentration) × 100 = Air Quality Index. The table below explains the different categories in the index.

Air Quality Index

Air Quality Index	Air Quality	Color Indicator	Health Advisory
0–50	Good	Green	No health advisory needed
51–100	Moderate	Yellow	People who are very highly sensitive to pollution should restrict heavy or prolonged activity
101–150	Unhealthy for Sensitive Groups	Orange	Older adults, children, and people with heart disease, asthma, or other respiratory ailments should reduce or restrict heavy physical activity
151–200	Unhealthy	Red	All people should restrict strenuous physical activity, and those who are in unhealthy and sensitive groups should avoid such activity entirely
201–300	Very Unhealthy	Purple	Everyone should avoid all physical activity
⩾301	Hazardous	Maroon	Can potentially have respiratory effects on all people, regardless of health; severe respiratory problems for those with asthma; heart and lung disease is aggravated

What is a **smog alert**?

A smog alert is a warning that air quality conditions are so bad outside that physical exertion outdoors may lead to respiratory discomfort. For people suffering from asthma, severe smog alert conditions could lead to asthma attacks and even hospitalization. An orange level or above reading in the Air Quality Index is cause for a smog alert to be issued.

How is **asthma** related to **air pollution** levels?

Asthma is rapidly becoming one of the major health issues in the United States, especially for children living in cities. A study released by the National Institute of Allergy and Infectious Diseases in 2008 showed that asthma particularly affected children in poor, inner-city areas and that car emissions of nitrogen dioxide, as well as air particulates and sulfur dioxide, were to blame. The study evaluated over 800 children in seven urban areas. These children had significantly higher rates of asthma and decreased lung function, which contributed to poor health and school absenteeism, even when the level of pollutants in the areas where they lived was deemed to be lower than acceptable standards set by the Environmental Protection Agency. Other studies have also shown that children have more allergies when they have been exposed to higher levels of air pollution.

What is the **Clean Air Act**?

In New York City in 1966, air pollution became so awful that hundreds of deaths were attributed to it that year. This and other serious pollution problems in America led to the 1970 Clean Air Act, which was later amended in 1977 and again in 1990. The Clean Air Act was designed to improve the air quality for all Americans. It was preceded by the Air Quality Act of 1967, which failed to do the job because it did not require environmental standards to be set. The Clean Air Act, however, charges the Environmental Protection Agency with setting standards regarding emissions of air pollutants (ranging from ozone and benzene to carbon monoxide and particulate matter) from factories, power plants, and all modes of transportation.

What are the **National Ambient Air Quality Standards**?

Part of the Clean Air Act, the National Ambient Air Quality Standards (NAAQS) are set values that are considered the maximum limits before air becomes too unhealthy to breathe safely. Industries that produce pollutants must adhere to these standards or else face fines and other penalties from the federal government. The standards are listed below.

National Ambient Air Quality Standards

Pollutant	Concentration	Over an Average Period
Carbon monoxide	9 parts per million 35 parts per million	8 hours 1 hour
Lead	1.5 micrograms per cubic meter	3 months
Nitrogen dioxide	100 micrograms per cubic meter	12 months
Ozone	120 parts per billion	hourly average can't exceed once per year over three years
Particulate matter	50 micrograms per cubic meter 150 micrograms per cubic meter	12 months 24 hours
Sulfur dioxide	80 micrograms per cubic meter 365 micrograms per cubic meter	12 months 24 hours

Who created the **first anti-air pollution law** in Western history?

England's King Edward I proclaimed in 1306 C.E. that coal burning was to be restricted while Parliament was in session. The penalty for violating this law was rather harsh by today's standards: death.

Methane gas emanated from livestock has actually become hazardous to the ozone layer because of the sheer numbers of cows, pigs, and other farm animals.

What caused the **Bhopal disaster**?

In December 1984, the U.S.-owned Union Carbide pesticide plant in Bhopal, India, leaked toxic chemicals (methyl isocyanate gas) that killed over 3,800 people. It was the worst industrial accident in history. Union Carbide paid a fine of $470 million to avoid facing criminal charges.

How much **pollution** is caused by **smoking tobacco**?

A 2004 study concluded that cigarette smoking released 10 times the amount of particulate pollution in the United States as was caused by diesel exhaust. Indoor pollution from smoking has been in the news a lot lately because the smoke is much more concentrated indoors, causing health problems even to those who refrain from the habit.

Is **air pollution destroying** our **architectural history**?

Yes. Many important buildings and monuments around the world are slowly being destroyed by air pollution and acid rain. The famous Taj Mahal in Agra, India, for instance, is turning from white to yellow because of pollution from car exhaust, so the local government has banned automobile traffic from coming closer than 1.25 miles (two kilometers). In other locales, the Sphinx in Egypt and the Parthenon in Greece are both being eaten away by acid rains. The acids are produced by sulfur dioxide mixing with water to form sulfuric acid solutions. This is particularly damaging to structures built from limestone and sandstone—common materials used by many civilizations throughout history—which are turned into powdery gypsum as a result. Over time, layers of building material crumble and turn to dust.

How do **factory farms** contribute to **air pollution**?

One of the debates currently raging between environmentalists and the agricultural industry concerns factory farms. These are large, usually corporate-owned facilities in which there are high concentrations of livestock or where vast areas of cropland contribute to fertilizer runoff. When it comes to air pollution, such farms are a major contributor to the problem. So much animal waste is generated by factory farms, that large lagoons filled with manure and urine have to be built to maintain all of it. Some of this liquid manure is dispersed by spraying it onto crops as fertilizer, but that doesn't really help the issue. These wastes generate large amounts of noxious pollutants, including ammonia gas, methane, and hydrogen sulfide, as well as carbon dioxide. Such gases contribute to acid rain and ozone depletion. In addition, livestock flatulence (to put it bluntly, burping and farting cows) is another source of methane gas. According to the Environmental Protection Agency, 20 percent of methane gas generated by human civilization comes from agricultural activities.

What did the **Environmental Protection Agency** (EPA) recently say about **power plant emissions**?

In 2004, the EPA estimated that every year about 2,800 people die from lung cancer and another 38,200 have heart attacks as a direct result of pollutants from power plants.

WATER POLLUTION

What is **acid rain**?

Motor vehicles and industrial activity release tons of pollutants into the air. When mixed together, the pollutants form sulfuric and nitric acids that later fall to the ground in rain or snow. This precipitation is known as acid rain. Acid rain is responsible for damaging lakes by killing plant and animal life and for killing trees around the world. Canada has been especially hard hit by acid rain caused by industrial activities in the United States.

Who first described the phenomenon of **acid rain**?

Scottish chemist Robert Angus Smith (1817–1884), who was very interested in water pollution and other issues regarding the environment and public health, conducted research in which he discovered, in 1852, that air pollution was causing acid rain. Smith also helped found the discipline of chemical climatology and was the author of the influential 1872 book *Air and Rain.*

Acid rain destroyed trees in the Bavarian Forest, but pollution controls are helping the forest to recover somewhat.

What key event brought the issue of **acid rain** to **prominence**?

During the 1960s, fish began to die in Scandinavian lakes at an alarming rate. When investigated, it was found that the lake water had become so acidic as to be unlivable for many species. The source of the acid was European industrial emissions creating acid rain and snow. Further research found that lakes in the United States, Canada, and other nations were suffering the same fate.

What is the **National Acid Precipitation Assessment Program (NAPAP)**?

Initiated by the Acid Precipitation Act of 1980, the NAPAP studied acid deposition throughout the United States. After 10 years, the researchers reported to Congress that, while acid rain was a problem, the danger was not quite as severe as initially feared. About four percent of the lakes in the United States were deemed unacceptably acidic, and 25 percent of these lakes had acidic water due to natural causes (for example, decaying vegetation debris can raise water acidity levels).

How **clean** is **rain water**?

Rain water—at least, when it is not acid rain—is normally very pure, except for the small particle around which the raindrop originally formed. Once the raindrop hits the ground and is absorbed into the soil or evaporates, it leaves behind this nucleus, which is often formed from such matter as a dust spec or tiny grain of salt.

What is the **difference** between **normal rain pH** and **acid rain pH**?

The normal pH of rain water is 5.0 to 5.6, while acid rain has a pH of about 4.3 (compared to distilled water, with a neutral pH of 7.0). Natural rain water is slightly acidic because of dissolved carbon dioxide, which actually makes it similar to club soda without the bubbles. Under some conditions, such as after dust storms, rain can become more alkaline, and volcanoes can add sulfur to clouds and contribute to acid rain. One extreme example of the latter happened in 1783, when Iceland's Laki volcano erupted and spewed so much sulfur into the air that acid rain killed the island's crops and polluted the air over Europe.

Does **rain** come in any **other colors** or is it always clear?

Yes, rain showers have occurred consisting of yellow or even reddish droplets. This usually happens when dust storms have seeded clouds heavily with nuclei rich in iron or other minerals. Yellow rain has also been observed as the result of pollens entering into clouds and then being absorbed into raindrops.

What is **eutrophication**?

Eutrophication is the excessive build up of nutrients in lakes, ponds, and other water bodies as a result of river and stream runoff containing pollution from fertilizer, sewage, and other waste products. While more nutrients in the water might at first sound like a good thing, it actually has harmful side effects, such as algae blooms that deplete water of oxygen and kill off aquatic species and other wildlife. Factory farms and private residents who fertilize their lawns are major culprits of this type of pollution. If you live by a fresh water source, such as a pond or stream, you can lessen the negative effects of runoff by creating a natural plant barrier between your lawn and the water. Among the plants that make ideal buffers are willow trees, birch, green ash, red maples, buttonbush, spice bush, some dogwood species, water oak and pin, sycamore, and smooth alder. These and other plants absorb the excess nutrients while also preventing erosion.

What is a **London killer fog**?

Despite its history of air pollution dating back centuries, Londoners seemed slow to learn from their mistakes. Air pollution from coal burning continued into the 1960s. The sulfur dioxide combined with London's famous fog, with the result being acid fog. In 1952, the thick fog became so dense that people could not see to walk or drive. Influenza, bronchitis, and pneumonia cases skyrocketed, and about 4,000 people died and another 100,000 were sickened that year from illnesses related to this killer fog.

What is **acid fog**?

Acid fog is just like acid rain. When sulfur dioxide is present in the air, it can be captured by water vapors of all sorts, including fog. If this weren't bad enough, acid fog tends to be much more acidic than acid rain—up to 10 times as acidic! This acid

Farmland irrigation is responsible for redistributing vast amounts of water to the point that agriculture is affecting rain patterns.

fog, which can linger in the air for hours or even days, can be as bad as walking through clouds of floating vinegar and is very destructive to plant life, as well as building materials ranging from iron to concrete.

How have **dams affected** the planet's **water cycle** in a significant way?

Currently, there are about 40,000 large dams (dams higher than 15 feet [5 meters]) in the world, with 19,000 of those being in China and about 5,500 in the United States. By erecting dams that block effectively 15 percent of the fresh water flow on the planet's surface, people have changed the Earth's water cycle significantly. Water cycle changes can, in turn, change such things as temperature and cloud formation. Naturally, the bigger the dam, and the reservoir created behind it, the greater the effect. China's Three Gorges Dam, located on the Yangtze River, is a huge dam that creates 20 times the kilowatts as America's famous Hoover Dam. The reservoir created by Three Gorges contains over 5 trillion gallons (19 trillion liters) of water and is over 400 square miles (1,036 square kilometers) in surface area. Scientists have observed that the climate around the dam has cooled and that rainfall has also changed.

Does **irrigation** of farmland affect the **climate**?

Extensive land irrigation can, indeed, affect the weather. As of 2000, the total amount of land on Earth that has been irrigated is about 689 million acres. About 60 percent of the world's fresh water supplies are used toward this purpose, representing 137 billion gallons (518.5 billion liters) of water daily. As one might imagine, when all of this water is spread over cropland, a certain amount of it will evaporate into the atmosphere. Hydrologists believe that this amount of evaporation is enough to increase the number of rainstorms that would otherwise occur.

RADIATION

What is **nuclear winter**?

A nuclear winter is what would follow a large-scale nuclear war. Radioactive particles, dust, and smoke released into the atmosphere would create a large cloud over the planet, blocking out sunlight and reducing temperatures worldwide. Plants and animals would die due to the extremely low temperatures and reduction of sunlight. An extended nuclear winter could cause the death of millions of people from starvation, cold, and other problems.

We haven't had a nuclear war (yet), so **what sources** are creating **radiation pollution**?

Man-made radiation in the atmosphere comes from primarily two sources: nuclear weapons testing and leakage from nuclear reactors, the latter mostly a result of nuclear plant accidents. After the United States invented the atomic and hydrogen bombs, there was extensive testing from 1945 through 1968. Over three hundred warheads were detonated during that time, mostly in desert regions and on small Pacific islands. The result was huge quantities of radioactive isotopes being spewed into the air, including carbon-14, strontium-90, iodine-131, and cesium-137. While precautions were taken by the military so that no one was killed in the initial blasts, radioactivity in the air traveled on wind currents and poisoned areas hundreds of miles from the tests. For instance, two days after a May 1953 test in Nevada, radioactive hail—some stones the size of tennis balls—fell in Washington, D.C. Later, the United States tested nuclear weapons underground in an effort to curtail this air pollution, but, of course, the radioactive wastes of subterranean nuclear explosions can easily make their way into underground water supplies. Other nations, too, have conducted nuclear weapons tests over the years, contributing to the problem.

What is **radon**?

Radon is actually a naturally occurring radioactive gas that is emitted from ground sources of decaying uranium and radium. It only becomes a dangerous hazard within manmade structures, such as private homes, where the radon can seep into basements and build up to harmful, carcinogenic concentrations. Lung cancer is one of the primary potential threats of radon poisoning. The government standard for "safe" levels of radon is under four picocuries per liter, which is roughly the equivalent of smoking a half pack of cigarettes per day. Because it is impossible to say which homes are more at risk for radon pollution than others, it is best to install

While safety technology has improved significantly at nuclear power plants throughout the country, the radiation leak at Pennsylvania's Three Mile Island back in 1979 remains in people's memories as a reminder of the risks to the environment.

radon detectors in the home. If a problem is detected, it can usually be solved by simple basement repairs or improving ventilation systems.

What was the world's **worst nuclear disaster**?

In April 1986, the Chernobyl nuclear power plant in the Ukraine, near the border with Belarus, had a major accident that released radiation into the atmosphere. The protective covering of the nuclear reactor exploded and deadly radiation escaped, immediately killing at least 28 people and giving 240 others radiation sickness, 19 of whom later died as a result. The radiation exposure that initially occurred is still killing people through related diseases, especially thyroid cancer, and this will continue for many years. More than 100,000 people were evacuated from the region, and deaths due to radiation poisoning continue as radioactive isotopes spread across Europe. The radiation cloud that resulted from the disaster traveled more than 1,300 miles (2,000 kilometers), infecting crops and livestock that then became unsafe to eat.

What happened at **Three Mile Island**?

Three Mile Island, Pennsylvania, was the site of the United States's worst nuclear accident. Luckily, no radiation was released into the environment and no one was killed. In March, 1979, the nuclear reactor at the Three Mile Island plant overheated, breaking the radioactive rods. Pennsylvania's governor recommended a voluntary evacuation of pregnant women and preschool children who lived within five miles (eight kilometers) of the plant. It was the unexpected self-evacuation of residents in the area that created major problems. The evacuations yielded surprising information about the lack of preparedness of communities for such an event, and have led to increased planning and preparedness for nuclear accidents and evacuations.

CLIMATE BASICS

What is the **difference** between **climate and weather**?

Climate is the long-term average weather for a particular place. The weather is the current condition of the atmosphere. So, the weather in Barrow, Alaska, might be a warm 70°F (21°C), but its tundra climate is generally polar-like and cold.

How are different types of **climates classified**?

The German-born, Russian climatologist Wladimir Köppen (1846–1940) developed a climate classification system that is still used today, albeit with some modifications. He classified climates into six categories: tropical humid, dry, mid-latitude, severe mid-latitude, polar, and highland. He also created sub-categories for five of these classifications. His climate map is often found in geography texts and atlases. In 1931, American geographer and climatologist Charles Warren Thornthwaite (1899–1963) published *The Climates of North America: According to a New Classification,* which takes into more thorough consideration how differences in geography affect local climates.

What was **Hermann Flohn**'s contribution to **climatology**?

German meteorologist Hermann Flohn (1912–1997) studied climate change on a macro scale. He conducted research on how large-scale changes in the Earth's climate—specifically, how the entire atmosphere circulates—affect the environment. He was also one of the first to advance theories on how human beings affect the climate.

How did the Inca civilization experiment with climate?

In the Urubamba valley in Peru, in a city called Moray, is the remains of a great amphitheater-like terrace system. Archaeologists and scientists now believe that this was a great agricultural laboratory, where each area of the terrace exhibited completely different climates, allowing the Incas to experiment with different climates and growing techniques.

What are **microclimates**?

Microclimates are small-scale regions where the average weather conditions are measurably different from the larger, surrounding region. Differences in temperature, precipitation, wind, or cloud cover can produce microclimates. Frequent causes of microclimates are differences in elevation, mountains that alter wind patterns, shorelines, and man-made structures that can alter wind patterns.

What do **butterflies** have to do with **chaos theory**?

American mathematician and meteorologist Edward Norton Lorenz (1917–2008) came up with chaos theory as a way of explaining the unpredictable way in which mathematical and natural systems (including weather) behave. The idea was that even the tiniest changes in initial conditions of a complex, dynamic system can lead to huge, measurable effects over time. As a colorful metaphorical illustration of this concept, he posed that a single butterfly flapping its wings in Brazil could be the instigator of a tornado in Texas. He called this the "butterfly effect."

What is the **National Climatic Data Center** (NCDC)?

Part of the National Oceanic and Atmospheric Administration, the NCDC serves as a vast archive of meteorological data called the World Data Center for Meteorology, which it then provides to various agencies, organizations, publications, insurance companies, and law firms all over the planet. Records go back as far as the nineteenth century, and include data ranging from modern radar and weather balloon reports to observations made by ships over a century ago. The meteorology data center is located in Asheville, North Carolina. The NCDC also runs the World Data Center for Paleoclimatology in Boulder, Colorado.

Is our **climate changing**?

Human lifespans, when compared to geologic time, are extremely brief, and so it is hard for us to imagine our planet's climate being much different than it is today. The fact is that over thousands and hundreds of thousands of years the Earth's climate has experienced wide fluctuations, ranging from the "snowball Earth" of about 635 million years ago to the extremely warm conditions (18°F, or about 8°C

How is climate change affecting geyser activity?

Geyser activity is affected by rainfall levels and the frequency of earthquakes. In Yellowstone National Park, which is home to about half of the world's 1,000 known geysers, rainfall feeds into the Madison River, which supplies geysers in the park with their water reserves. Less rainfall in times of drought translates into less pressure in the geysers' reservoirs, and leads to fewer eruptions. In a recent study of rainfall and climate in Yellowstone, it was found that the amount of water flowing down the Madison River has decreased by about 15 percent from the years 1998 through 2006.

This decrease in rainfall has been blamed on global warming, which has affected precipitation in the surrounding states of Wyoming, Montana, and Idaho. The study associated lower rainfalls with an increase in the intervals between eruptions of geysers. For example, Old Faithful used to erupt every 75 minutes, and in 2006 it was erupting every 91 minutes.

warmer, on average, than today) of a 100 million years ago when dinosaurs were flourishing. Over the millennia, there have been many periods of hot and cold. In the past, these have been caused by a variety of factors, ranging from volcanic activity to plate tectonics to the possibility of an asteroid crashing into the planet. Today, scientists fear we are undergoing a new and dramatic climactic change with one important difference: they believe humans are causing it.

What is **paleoclimatology**?

Paleoclimatology is a fascinating discipline that combines paleontology with climate studies. Understanding what the Earth's climate was like millions of years ago can prove very valuable in understanding today's weather. In addition, it's just plain cool to find out facts such as that dinosaurs once roamed Antarctica, tropical fruits used to grow in Oregon, and just 8,500 years ago the temperature in Greenland was about 10°F (5°C) warmer than it is today. Paleoclimatologists discover this information by examining animal and plant fossils, conducting ice core studies, and examining soils and rocks buried deep underground. Clues may be found in some of the most unlikely places. For instance, when ancient pine needles hoarded by rats 30,000 years ago were discovered, the composition of the plant matter was analyzed to learn what carbon dioxide levels were like back then.

Who is considered one of the most **important pioneers** in the study of **climate change**?

English meteorologist and climatologist Hubert Horace Lamb (1913–1997), who founded the Climatic Research Unit at the University of East Anglia in 1971, is considered by many to be the greatest climatologist of the twentieth century. He began his career as a weather forecaster for the Irish Meteorological Office and later was

The Strokkur Geyser erupts in Iceland. Some scientists theorize that changing rainfall patters affect the frequency of geyser eruptions.

with the British Meteorological Office. While there, he participated in a Norwegian whaling expedition in Antarctica from 1946 to 1947. It was during this time in Antarctica that he began to study how the world's climate must have changed, and he pursued this subject after joining the Climatology Division of Britain's Meteorological Office in 1954. Using records there, he researched and published papers and books about how Great Britain's climate had changed noticeably since the middle of the nineteenth century.

What is the **Milankovič Cycle**?

Earth's orbit is currently fairly circular around the Sun. However, this is not always the case. Our planet experiences shifts from the current circular orbit to a much more elliptical one during which the difference between perihelion and aphelion is very marked. The entire cycle—from circular to elliptical and back again—takes about 95,000 years to complete. During the periods when the Earth is in its more elliptical orbit and ventures farther from the Sun, ice ages tend to occur. The theory was first formulated by Serbian geophysicist Milutin Milanković (also spelled Milankovitch, 1879–1958), who was well known for his research into ice ages. His theories were later verified, in 1976, through the study of sediment cores taken during deep-sea explorations.

Who else formed theories that **ice ages occur in cycles**?

Before Milanković , French mathematician Joseph Adhemar (1797–1862) published his book *Revolutions of the Sea* in 1842. In it, he suggested ice ages occur in 22,000-year cycles that matched the precession of the equinoxes. Scottish geologist James Croll (1821–1890) later elaborated on this theory. However, scientists did not know enough about the history of ice ages to compare this theory with actual data, and so the hypothesis languished for another century.

Why does the **Earth's orbit change** from circular to elliptical over time?

Like all the planets in our solar system, our world is subjected to the gravitational tugs of not only the Sun, but all the other planets as well. We know that our Moon causes tides and other gravitational effects that we can see every day, but Earth's orbit is also influenced by the gas giants Jupiter and Saturn. These planets are big enough to pull Earth's orbit out of shape as they circle the Sun; then the Sun's gravity eventually pulls it back again in an extremely slow tug of war.

What is a **nuclear winter**?

Scientists Richard Turco and Carl Sagan made the world cognizant of the effects of a global nuclear war in the early 1980s, and Sagan published a popular book on the subject, *The Nuclear Winter,* in 1983. While most people were already terrified of the nuclear arms race that had been going on between the United States and the Soviet Union since the 1950s, the idea that ever major city on Earth could be obliterated by atomic and hydrogen bombs was just the beginning. Sagan and Turco showed that it wouldn't even take all of the then-50,000 nuclear warheads to kick up enough dust and debris to block out the Sun's warmth. Clouds of irradiated dust would be blown into the stratosphere, where they would circle the planet for months, plunging us into an artificial winter that would destroy crops and lead to a global famine.

Since the publication of *The Nuclear Winter,* many scientists have come to believe a similar scenario could happen if a large asteroid hit the planet. Indeed, this is one theory about how the dinosaurs may have become extinct 65 million years ago. Another possible cause of a nuclear winter would be planet-wide volcanic eruptions, which some scientists theorize led to a "Snowball Earth" hundreds of millions of years ago. Actually, when volcanoes are involved it would perhaps be more accurate to call these "volcanic winters."

How did **Mikhail Ivanovich Budyko** calculate the temperature of Earth's climate?

Mikhail Ivanovich Budyko (1920–2001) was a Belorussian meteorologist and physicist, as well as a pioneer in the specialized field of physical climatology. He was the author of the 1956 book *Heat Balance of the Earth's Surface.* In this work, he used the principles of physics to explain, as no one had done before, how energy from the Sun is absorbed and then radiated back into the atmosphere by the Earth. Budyko's research led him to become concerned about climate change, and he was one of the first to realize that a build-up of carbon dioxide created by human industry was causing the atmosphere to warm up. He predicted that by 2070 the average temperature on Earth would be 6.3°F (3.5°C) higher than 1950 temperatures. Budyko was also one of the first to predict that a nuclear winter would result from a nuclear war.

Where can I find reliable **climate statistics online**?

There are many websites about the weather available on the Internet; some are better than others. You can get lots of information from government websites, such as the National Weather Service at http://weather.gov and the National Oceanic and Atmospheric Administration at http://www.noaa.gov. There are also some online databases available. WorldClimate at http://www.worldclimate.com is a searchable online database that includes more than 85,000 climate statistics from all over the world. By typing in the name of a city, you can obtain data concerning rainfall and temperature. Weatherbase at http://www.weatherbase.com has information on nearly 16,500 cities around the world; it even lets you pick whether you want to see statistics in metric or U.S. units.

ICE AGES

What is an **ice age**?

Geologists define an ice age as a period when more of the Earth's surface was covered by larger ice sheets than in modern times, or when cool temperatures endured for extended periods of time, allowing the polar ice to advance into lower, more temperate latitudes. The last ice age event, which ended about 12,000 years ago, is referred to as the "Ice Age" or the "Great Ice Age"—both usually capitalized—a time when ice covered nearly 32 percent of the land and 30 percent of the oceans.

Who **first proposed** the idea of **ice ages**?

Several scientists over the centuries came close to proposing the idea of ice ages. Scottish naturalist James Hutton (1726–1797) observed strangely shaped glacial boulders (erratics) near Geneva, Switzerland. Based on this, he published his theory in 1795 that alpine glaciers were more extensive in the past. In 1824, Jens Esmark (1763–1839) proposed that past glaciation had occurred on a much larger continental scale.

Geologist Louis Agassiz was among the first to form a theory about past ice ages. (*NOAA*)

But the most persuasive argument for ice ages came in 1837, when Swiss-American geologist Louis Agassiz (1807–1873) gave his now-famous speech on past widespread ice age conditions. He proposed that nearly all of northern Europe and Britain had once been covered by ice, and he subsequently found evidence for his theory in New England. Others eventually uncovered additional evidence. In 1839, United States geologist Timothy Conrad (1803–1877) discovered evidence of polished rocks, striations, and erratic boulders in western New York, supporting Agassiz's theory that Ice Age glaciation was worldwide. In 1842, the first attempt to explain the ice ages using an astronomical connection was made by

French scientist Joseph Adhemar (1797–1862). He proposed that the ice ages were the result of the 22,000-year precession of the equinox, a natural movement of the Earth's axis that causes the seasons to switch over thousands of years. In other words, the current summer months would become winter months and vice versa.

What **causes ice ages**?

No one knows why ice ages occur, but there are several theories. One possibility is that the Sun's energy varies in intensity over time. Each time there is a decrease in activity, an ice age may occur as the Earth cools. Another possibility is an increase in dust in the atmosphere, either from volcanoes or a large meteorite impact. The debris from either event would reflect more of the Sun's light into space (albedo), cooling down the atmosphere and causing more snow and ice to form. This would also further increase the world's albedo, as even more sunlight would reflect off the ice and snow. However, this theory has a problem, as many of the other theories do, in that it doesn't explain what causes the ice sheets to retreat.

What were the **major glacial ice ages** over time?

Evidence from the geologic record shows that ice ages have occurred relatively few times in Earth's history, with the first known large-scale ice age taking place approximately 2.3 billion years ago. (In the last about 670 million years, ice ages have occurred less than 1 percent of the time.) Geologists believe there have been five major ice ages over geologic time. The following lists the occurrences:

- 1.7 to 2.3 billion years ago (Huronian Era, Precambrian)
- 850 million years ago (Cryogenian Period)
- 670 million years ago (Proterozoic Era, Precambrian)
- 420 million years ago (Paleozoic Era, between the Ordovician and Silurian Periods)
- 290 million years ago (Paleozoic Era, between late Carboniferous and early Permian Periods)
- 1.7 million years ago (Cenozoic Era, Quaternary Period, Pleistocene epoch)

What was **Snowball Earth**?

Scientists now believe that at between two to four times between 850 and 580 million years ago, the planet was covered completely in ice and early life forms were nearly wiped out: a "Snowball Earth." There are a couple of theories as to why this occurred: one being that the Sun was about six percent cooler than it is today; the other is that the continents had all traveled—due to plate tectonics—mostly south; ocean currents traveled easily around the planet without interruption, and volcanic activity dwindled to a minimum. The result was that significantly less carbon dioxide was being generated and expelled into the atmosphere. In addition, the Earth's axis was significantly more tilted at the time (by 54 degrees versus today's 23.5 degrees), causing more drastic seasonal extremes.

Once a period of cooling began and ice shelves formed, more of the Sun's light was reflected, thus causing the cooling cycle to accelerate more and more until everything was frozen. Early theories about this, proposed by such scientists as George Williams at the University of Adelaide and Joe Kirschvink at the California Institute of Technology, include the effects of plate tectonics as causing massive volcanic eruptions worldwide. This would lead to an extended winter of unprecedented scale that turned Earth into one big frozen lump in space that would last for millions of years. Nearly all life was wiped out. In fact, one matter of debate concerning this theory was that critics felt such a Snowball Earth would have extinguished all life. This notion was laid to rest in the 1990s when life was found to thrive near geothermal vents deep under the ocean.

The Snowball Earth cycle was only broken because volcanic action and plate tectonics would cause carbon dioxide levels to build under the ice, especially since there would be no liquid water to dissolve minerals or aid in the dissipation of CO_2 levels. The result would be an inevitable, sudden massive release of CO_2 that would lead to a comparatively brief period of extreme heat with temperatures averaging 120°F (50°C). However, the thaw retreated to another snowball period a couple more times over millions of years before the continents moved into positions that created a more stable geophysical state of the planet. Volcanic activity moderated, as did carbon dioxide levels. Another Snowball Earth could happen again, however, as the continents, hundreds of millions of years from now, drift back together to form a new supercontinent.

What were the **glacial and interglacial periods** in the last Ice Age?

In the last Ice Age (and in all ice ages), there were cycles of glacial (when ice covered the land) and interglacial (relatively warmer temperatures) times. Corresponding with these times, the glaciers advanced or retreated. Scientists believe the last Ice Age—also called the Pleistocene Ice Age—had eight cycles. The following lists these stages for North America (stage names for northern and central Europe differ). All dates are approximate:

Glacial/Interglacial Periods

Approximate Years Ago	North American Stage
75,000–10,000	Wisconsin*
120,000–75,000	Sangomonian (interglacial)
170,000–120,000	Illinoian
230,000–170,000	Yarmouth (interglacial)
480,000–230,000	Kansan
600,000–480,000	Aftonian (interglacial)
800,000–600,000	Nebraskan
1,600,000–800,000	Pre-Nebraskan

* Note: During the Wisconsin glacial stage, an interstadial period occurred—a time not warm or prolonged enough to be called an interglacial period.

How did the **last Ice Age** begin and end?

Approximately 1.7 million years ago (the beginning of the Quaternary Period, Pleistocene epoch) geologists believe the plains of North America cooled. As a result, large ice sheets began to advance south from the Hudson Bay area of Canada and eastward from the Rocky Mountains. These ice sheets advanced and retreated many times toward the end of the Pleistocene epoch in intervals lasting from 10,000 to 100,000 years. This most recent ice age ended about 10,000 years ago, when the ice retreated to its present polar positions. Currently, the Earth is nearing the end of an interglacial (warmer) period, meaning that another ice age might be due in a few thousand years.

What are **major and minor ice age periods**?

Of course, not all scientists agree on how to divide the periodic ice ages, with some calling for a more strict division between the times and temperatures. For example, some scientists believe a major ice age period should be defined as lasting about 100,000 years, with a 9°F (5°C) decrease in temperature between glacial and interglacial periods; a minor ice age period, lasting about 12,000 years with a 5°F (2.8°C) decrease in temperature; and a smaller ice age period, lasting about 1,000 years with a 3°F (1.7°C) decrease in temperature.

How much of the **Earth was covered in ice** during the various ice ages?

Because of extensive erosion, it is difficult to determine the extent of ice sheets during the various ice ages. But scientists do know some things about the last ice age—the Pleistocene Ice Age—a time when up to 10 percent of the Earth was covered (although not simultaneously) by often miles-high ice. At their greatest extent, the Northern Hemisphere glaciers and ice sheets covered most of Canada, all of New England, much of the upper Midwest, large areas of Alaska, most of Greenland, Iceland, Svalbard, and other arctic islands, Scandinavia, much of Great Britain and Ireland, and the northwestern part of the former Soviet Union. In the Southern Hemisphere, glaciers were much smaller, with the main effects being cooler and much drier weather.

During the last stage of the Ice Age (the Wisconsin stage in the United States), ice sheets covered parts of Eurasia and much of North America, extending as far south as Pennsylvania. As the climate warmed up, scientists estimate that sea level rose about 410 feet (125 meters), an average rate of an inch (2.5 centimeters) per year for roughly 5,000 years. Interestingly, although most of the huge northern ice sheets melted, the Antarctic ice sheet decreased by only 10 percent.

What was the **"Little Ice Age"**?

The "Little Ice Age" was an interval of relative cold, beginning about 1450 and lasting until about 1890 (the coldest periods of 1450 and 1700 are often divided into the two Little Ice Ages). It occurred during the current warm, interglacial period, but is not considered a full glacial episode, since the high latitudes of the Northern Hemisphere landmasses remained largely free of permanent ice cover.

Even so, much of the world experienced cooler temperatures during this time of at least 2°F (1°C) lower worldwide average surface temperatures. It was a time of renewed glacial advance in Europe, Asia, and North America, with sea ice causing havoc in the colonies of Greenland and Iceland. In England, the Thames River froze; in France, bishops tried to halt glacial advances with prayer. Several historians also believe the low temperatures caused social conflict and poor food production. Thus, this may have been partially responsible for war and hunger during that time.

Just like the major ice ages, no one really knows what caused the Little Ice Age, though English astronomer Edward Maunder (1851–1928) first hypothesized that it had something to do with solar activity. Other scientists attribute the cooling down to volcanic eruptions, changes in the ocean circulation, changes in the Earth's orbit, the wobbling of the Earth's axis, or even our planet's passage through clouds of interstellar dust.

Will there be an **ice age in the future**?

Yes, eventually the Earth will again cool and ice will cover land at higher latitudes and elevations. When this might happen, no one is really sure.

GLOBAL WARMING

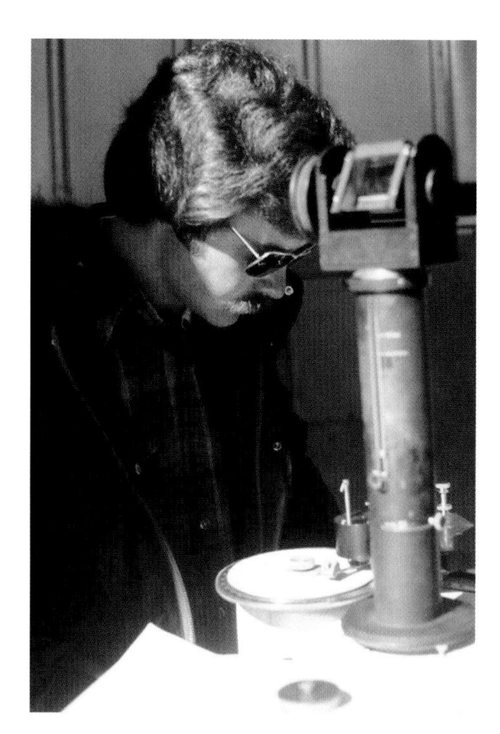

A scientist compares ultraviolet wavelengths at a Mauna Loa, Hawaii, research station using an ozone spectrophotometer. (*photo by John Bortniak, NOAA Corps*)

What is the **greenhouse effect**?

The greenhouse effect is a natural process of the atmosphere that traps some of the Sun's heat near the Earth. The problem with the greenhouse effect, however, is that it has been unnaturally increased, causing more heat to be trapped and the temperature on the planet to rise. The gases that have caused the greenhouse effect were added to the atmosphere as a byproduct of human activities, such as combustion from automobiles.

Who **first formed the theory** about the **greenhouse effect**?

Irish physicist, mathematician, and chemist John Tyndall (1820–1893), who succeeded Michael Faraday (1791–1867) as the superintendent of Britain's Royal Institution, began conducting research in radiant heat in 1859. He soon concluded that water vapor was vital for holding in warmth in the Earth's atmosphere, and that other gases, such as carbon dioxide and ozone, also played a role. He proceeded to play with a number of calculations, changing the amounts of these gases in his formulas to discover what the results would be. Tyndall concluded that increasing a gas like carbon dioxide would have significant effects on the climate that we now call global warming.

Is the **greenhouse effect** a **good thing** for life on Earth, or **bad** for the environment?

Like so many things about life on Earth, moderation is the key. Some greenhouse effect is a very good thing for life on Earth. Without any such effect on Earth, the oceans would eventually freeze. If the greenhouse effect increases significantly, however, many living organisms and species, as well as environmental systems that have developed over a long time—including human civilization—will face substantial challenges, and possibly even extinction. In the most extreme case, a runaway greenhouse effect like that on the planet Venus, would cause all life on Earth as we know it to cease to exist. On the other hand, many plants will benefit from increases in carbon dioxide levels, warmer temperatures, and longer growing seasons.

What is **global warming**?

Global warming and the greenhouse effect are not necessarily the same thing. While the greenhouse effect can cause global warming, other things can lead to the planet warming or cooling. Other factors include changes in geography (plate tectonics) and cycles of the Earth's orbit around the Sun.

It is known that global warming is necessary to a certain point: Without the ability of certain gases in the Earth's atmosphere to help retain radiation from the Sun, our planet would be a cold ice ball in space. These gases act like the glass of a greenhouse (thus the name greenhouse gases), trapping much of the energy emitted from or bounced off the ground and keeping our world warm. This, in turn, allows organisms—plants, animals, and otherwise—to live.

More recently, global warming has been used to describe the unnatural increase in the average surface temperatures around the world. Many scientists (and others) believe humans have pumped excess amounts of greenhouse gases, such as carbon dioxide, methane, and nitrogen oxides, into the atmosphere, causing temperatures to rise. Something has already raised the surface temperatures about 0.5°C (1°F) in the past 100 years, and scientists believe it to be human-induced.

What is the **global average temperature**?

Data on temperatures for the twentieth century, which includes both land and ocean surface temperatures, averaged to 53.6°F (12°C). A 2007 assessment showed that, during the early years of the twenty-first century, average temperatures were already up 1.28°F (0.71°C) compared to the previous century averages. This data includes a very warm year in 2002 (though 1998 was actually the warmest year since 1990) , and the fact that, taken by themselves, land-surface temperatures have been particularly warmer: 3.4°F (1.89°C). Ocean surface temperatures in 2007 were the fourth warmest in a record database spanning 128 years.

What **gases and chemicals** are considered to have the **greatest effect on global warming**?

Anything that increases carbon dioxide (CO_2) levels is a risk factor in heating up the planet. In addition to CO_2, methane gas (CH_4) and chlorofluorocarbons (CFCs) are culprits in global warming. Methane (from livestock, coal mining, and also natural sources like peat bogs and termites decomposing wood) is actually 25 times more effective in holding heat within our atmosphere, and CFCs are 20 thousand times more efficient than CO_2.

Methane gases during the 1990s were rising at an annual rate of about 0.8 percent; from 1997 to 2007, however, methane levels stabilized, and scientists believed that a balance had been reached between production of methane gas and its rate of dissipation in the atmosphere. In 2007, however, a sudden spike in methane levels was noted, especially in the Northern Hemisphere, for reasons not well understood. Theories include the idea that global warming increased methane production from wetlands bacteria (i.e., the result of melting permafrost, especially in Siberia), or that the amount of OH (hydroxyl free radical), which breaks down methane, is decreasing in the atmosphere. Today, the amount of methane in the air is more than twice what it was before the Industrial Revolution (about 1,775 parts per billion versus about 700 parts per billion). The current rate of increase for methane is about 10 parts per billion annually, which is considered a significant spike. Meanwhile, because of government regulations, CFC levels have been steadily decreasing in the atmosphere, which is particularly good news for the ozone layer.

Is **carbon dioxide** the most **harmful global warming gas**?

No. Actually, water vapor contributes much more to global warming than CO_2 or methane do. The problem is that human activities often increase particulates in the air, which provides nuclei for precipitation to form in clouds.

Who first hypothesized that the amount of **carbon dioxide** in the atmosphere had something to do with **climate change**?

Swedish chemist Svante Arrhenius (1859–1927) was the first to propose that the atmosphere would hold more heat as concentrations of carbon dioxide increased.

Does **all carbon dioxide** released as pollution **remain in the atmosphere**?

No, not at all. A lot of it is reabsorbed by plants, as well as the oceans, where it turns into carbonic acid. Forests and oceans can only do so much, however, especially when people are cutting down forests all over the planet. It only takes one acre of forest to absorb about 13 tons of gas and particle pollutants annually. What is not

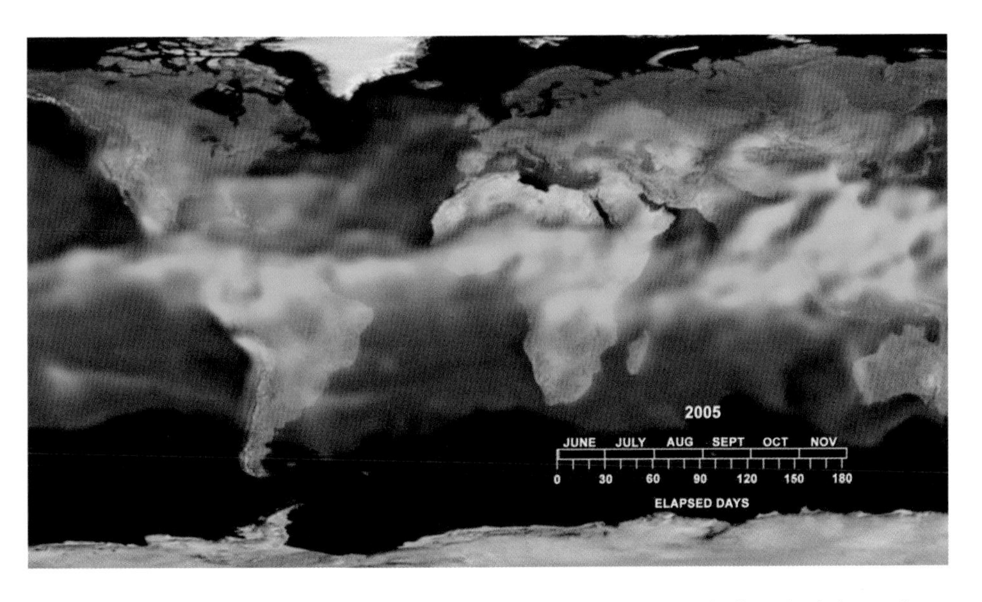

Water vapor concentrations and distribution in the atmosphere is a factor in global warming that has twice the impact of carbon dioxide levels.This satellite image depicts water vapor distribution during the fall of 2005, with the brighter shades of blue indicating water vapor at higher elevations. (*NASA/JPL*)

absorbed remains in the atmosphere, and people are producing so much CO_2 and other pollutants that there is a net gain in global warming gases.

How much carbon dioxide is being produced by cars and industrial emissions in the United States?

A 2004 study from the Environmental Protection Agency noted that total emissions from 1990 to 2004 have increased by 15.8 percent. While this is not great, when compared to an increase in the U.S. gross domestic product of 51 percent over the same period, it is a somewhat controlled increase.

How much **carbon dioxide** am I producing when I **drive my car or truck**?

According to the Environmental Protection Agency, burning a gallon of gasoline creates 19.4 pounds (8.8 kilograms) of carbon dioxide. Burning a gallon of diesel fuel creates 22.2 pounds (10 kilograms) of carbon dioxide. So, for example, if you have a 15-mile, one-way commute to work, work 250 days a year, and drive a gas-powered sedan that gets 18 miles per gallon, you would produce over 8,000 pounds (3,600 kilograms) of carbon dioxide pollution annually. Multiply that by the number of people driving every year worldwide, and you can see the problem!

Which **country** produces the most **greenhouse gas**?

This dubious honor used to belong to the United States. The economic boom in China, however, has led to a corresponding surge in pollution. While, per capita, Americans still cause more pollution, there are four times as many Chinese as Amer-

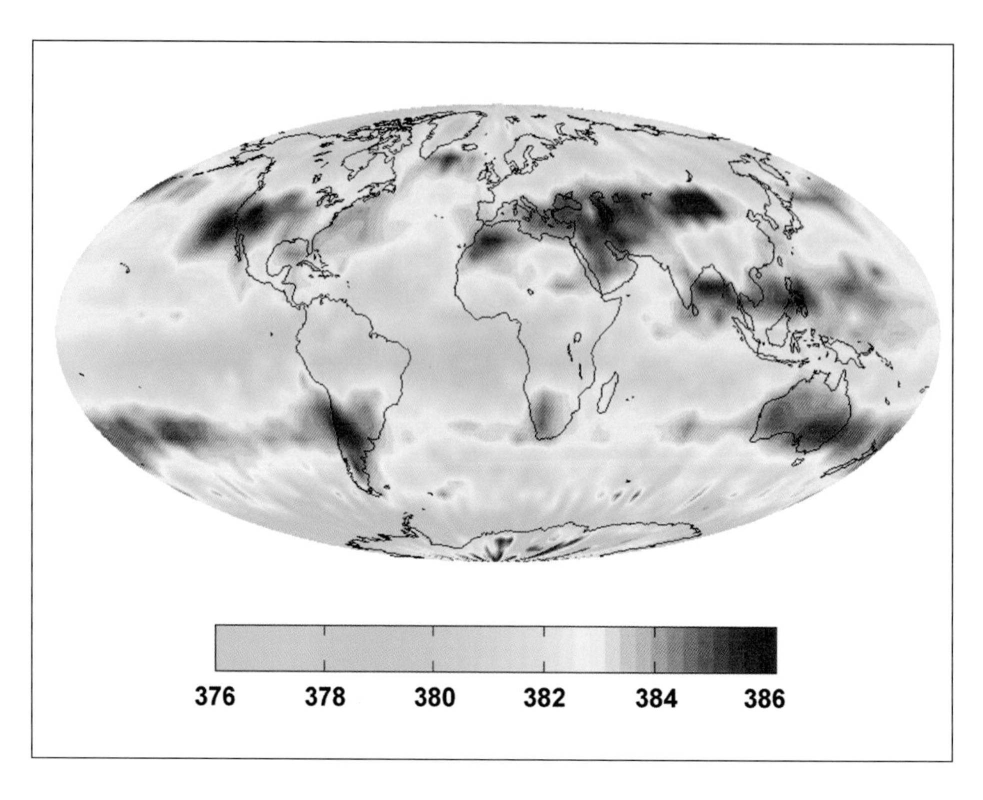

This July 2008 image was taken by the Atmospheric Infrared Sounder (AIRS) about NASA's Aqua spacecraft. It shows carbon dioxide distribution in parts per billion. (*NASA/JPL*)

icans. A 2007 study published by the Netherlands Environmental Agency noted that Americans produce about 20 tons (18 metric tons) of carbon dioxide emissions per capita, versus five tons for each Chinese citizen and about 10 tons (9 metric tons) for each European. Industrialized nations emitted about 16.1 tons (14.6 metric tons) of greenhouse gases per capita versus about 4.2 tons (3.8 metric tons) per capita for those living in developing countries. The study also noted that if the increase in greenhouse gases is not halted within 10 to 25 years the planet will experience a runaway greenhouse effect with average temperatures rising several degrees Celsius, causing ice caps to melt and coastal regions worldwide to become flooded.

How many **people** will find themselves in **flooded coastal regions** should **global warming melt the ice caps**?

Worldwide, over 150 million people would find themselves underwater by the year 2070 should scenarios about global warming play out. This includes 130 important port cities, such as New York City, Tokyo, Hong Kong, Mumbai, and Bangkok. A 2007 estimate said that this would have an economic impact of $35 trillion dollars, an amount no nation is currently equipped to surmount. Not only would such a flood destroy coastal cities, but mass migrations would occur, causing political and social unrest that many believe would inevitably lead to military conflicts, famines, and disease.

Is there a connection between **global warming and geology**?

Yes, there is an indirect connection between global warming and geology because changes in one part of the Earth's complex systems affect other parts. In particular, a change in the global atmosphere (and biosphere) can affect the rock cycle: A rise in sea level can change the expanse of glacial ice, change positions of deserts, and cause ocean waters to inundate coastlines—all of which would change rates and types of weathering taking place on our planet.

Another important connection involves dissolved carbon dioxide in the oceans and surface water. Undoubtedly, the gas is taken up by organisms, but it also precipitates out of the ocean and surface waters to form certain sedimentary rocks. Carbon dioxide then returns to the system from a multitude of places, including the dissolution of carbonate minerals in rocks and shells, weathering of carbonate minerals, volcanic eruptions or hot springs, reactions with the atmosphere, respiration of organisms, and through streams and groundwater. Most scientists agree that a major change in the amount of carbon dioxide in or out of the environment can affect us all. Not only would humans and other living organisms be affected, but also the natural cycles connected to the world's geology.

Still another geologic–global warming connection may be found in the weather and climate. If global warming continues, more powerful and intense weather systems may develop. Such events as superhurricanes would cause a great deal of erosion along coastlines, not to mention deluging countless rivers and creeks inland. In terms of climate change, variations in vegetation patterns could contribute to more erosion in various areas; glacial, polar, and sea ice would change drastically, altering the amount of radiation reflected back into space—thus enhancing the warming effect; and changes in the hydrologic cycle would alter stream flow and groundwater levels.

What is the **evidence supporting global warming**?

More and more scientists are becoming convinced that we are seeing an increasingly swift transition toward a warmer climate on Earth. Data is constantly being collected by meteorologists, environmentalists, and other scientists working in labs and out in the field. While there is no definitive proof that will assuage all doubters, the evidence is mounting. Below are some of the reasons why the fear of global warming is gaining ground.

- *Record-breaking temperatures*—Cities around the world have been recording high temperatures in the last decade or two that have exceeded all previous records.
- *Drought and rain pattern changes*—Dry areas are becoming drier, and wet areas are becoming wetter; floods are becoming more frequent, as rivers fill with record levels of water. For example, the American Midwest and Great Plains states have seen significant increases in flooding in recent years (though part of this can be blamed on farming and levee activities). The Northwest is

Global warming negatively impacts coral reefs, such as this reef near Chuuk, Micronesia. Warming waters kill off corals, causing a "bleaching" effect that destroys rich habitat. (*photo by Dwayne Meadows, NOAA/NMFS/OPR*)

also seeing much more rain, while the southwest and, especially, southeastern states have been experiencing extended droughts.

- *More extreme weather*—Some—though certainly not all—scientists feel that Hurricane and tornado activity appears to be on the rise, with the United States and Southeast Asia being particularly hard hit. One of the worst years of recent memory, 2005, included the devastating Hurricane Katrina that destroyed New Orleans and other Gulf Coast cities. Climatologists noted that there had been a lull in hurricanes during the 1970s and 1980s, but since 1995 there has been a significant upswing. Between 1995 and 2005, nine of the hurricane seasons had above normal incidents of storms.
- *Warming oceans, coral reef bleaching*—There is widespread and undisputed evidence that coral reefs around the globe are in trouble. Although coral reefs occur in tropical, warm waters, too much warming will kill coral; first, because the organisms simply don't tolerate such temperatures, and second because warmer water is not able to maintain as much dissolved calcium, which coral need to survive. When corals die, the vivid colors the tiny animals lend reefs disappear; hence the term "bleaching."
- *Melting glaciers*—The most commonly used piece of evidence for global warming has been the changes seen in glaciers, especially in the North, but also in Antarctica and on mountaintops around the world. Photographs have shown,

for example, that Greenland glaciers are retreating at a rate of about 100 feet (30 meters) per day. In Antarctica, the Larsen Ice Shelf, covering an area roughly the size of Rhode Island, broke off the mainland. Meanwhile, the glacier on top of Mount Kilimanjaro in Tanzania is quickly melting, and scientists speculate there will be no more snow on the mountain by 2020.

Is **global warming** causing the world's **glaciers to melt**?

Many scientists believe that greenhouse gases are directly causing glaciers in all parts of the world to melt and recede at an unprecedented rate. It is thought that by 2030 there will be no glaciers in Glacier National Park in Montana. In East Africa, Mt. Kenya's Lewis Glacier in Kenya has lost 40 percent of its size in just the last 25 years.

What is the **ICESat** mission?

ICESat (short for Ice, Cloud, and Land Elevation Satellite) is a satellite launched by NASA on January 12, 2003, to collect data on everything from land topography and vegetation to information on aerosol levels. On board is a single monitoring device called the Geoscience Laser Altimeter System (GLAS). One of the chief missions of ICESat, however, is to find out how the planet's ice sheets are changing.

What are the **consequences** when **glaciers melt**?

Glaciers that have melted in the Himalayas, home to the world's largest mountains, have filled up and burst the banks of nearby glacial lakes, filling rivers and causing widespread flooding and death to nearby populations downstream. Similar consequences will likely befall those now living near other glaciers around the world.

How did abrupt climate changes in the past come to be called **Dansgaard-Oeschger events**?

Danish geophysicist Willi Dansgaard (1922–) discovered that by measuring the levels of oxygen isotopes and deuterium (a hydrogen isotope) in glaciers, as well as dust content and the acidity of ice, it was possible to reconstruct what the Earth's climate was in the past. He came about his conclusion by studying ice cores drilled out of glaciers in Greenland in the 1960s with Swiss physicist Hans Oeschger (1927–1998), the inventor of the Oeschger counter, a radiation measuring device. Oeschger and Dansgaard drilled ice corps samples dating back some 150,000 years in Earth's history. Layers in the ice showed that there had, in that time, been 24 abrupt changes in the world's climate. These are now called Dansgaard-Oeschger events.

Does **global warming** increase the incidence of **diseases**?

Warmer temperatures tend to encourage pests to flourish. This includes everything from insects and rodents to viruses and germs. Mosquitoes, which are responsible

A worker drills for ice core samples in the Arctic Ocean. Such samples reveal important historical details about changes in the Earth's atmosphere. (*photo by Mike Dunn, North Carolina State Museum of Natural Sciences, courtesy NOAA*).

for killing more human beings than any other animal on the planet because they spread disease through exposure to blood, thrive nicely in hot to temperate climates. Their breeding cycles become shorter, which allows them to reproduce more frequently each year. Mosquitoes spread diseases such as malaria and dengue fever. Rodents, carrying disease-spreading parasites like fleas and ticks, also spread their ranges as climates warm. The problem is multiplied by the fact that along with global warming comes drought, flooding, and famine, which lead to poverty, homelessness, and unsanitary water supplies. All of these encourage unsanitary living conditions that help spread diseases, such as cholera. According to a 2008 statement by the World Health Organization, about 150,000 people are dying every year from malnutrition, diarrhea, and malaria, much of it provoked by climate change. Hardest hit are poorer countries in Africa and Southeast Asia.

What is the **effect of global warming** and climate change on **average temperatures**?

By the year 2100, relative to 1990, world temperatures could rise from 2 to 11.5°F (1.1 to 6.4°C) and sea levels may rise about three to five feet (one or two meters).

Do we have any **evidence of global warming** in the **past**?

Yes, there is a great deal of evidence for global warming events in the past. The following lists two of the more well-known ones:

Mid-Cretaceous Period—During this period (between about 120 and 90 million years ago), new ocean crust was produced at about twice the normal rate. Large volcanic plateaus were forming in the ocean basins, ocean temperatures were very high, and there was a peak in worldwide petroleum formation. Just as startling was the sea level, which was about 330 to 660 feet (100 to 200 meters) higher than at present. The reasons for the high temperatures were probably numerous, including the release of carbon dioxide (a greenhouse gas) by volcanic eruptions, creating a "supergreenhouse" effect. This led to temperatures about 20 to 22°F (10 to 12°C) above our current

How high would the oceans rise if all the ice sheets melted?

It is estimated that if all the ice sheets melted, including the polar ice caps, ice on Greenland and Iceland, and all the glaciers on the planet, the average worldwide sea level would rise by about 250 feet (76 meters). Such a rise would decimate almost all coastal cities, shrink the living space on all the continents, cause massive climate fluctuations, and change our lives forever.

average global temperatures. Interestingly enough, it is thought that the large volume of basalts that erupted on the ocean floor displaced a great deal of ocean water, causing sea levels to rise. And with the rise in sea level and temperatures, organisms flourished, eventually providing material necessary for petroleum formation.

Eocene Period—During this time (between about 55 to 38 million years ago), temperatures also increased, with tropical vegetation reaching about 45 to 55 degrees north and south of the equator, or about 15 degrees higher than today. Based on rock samples, it appears that the Earth had between 2 and 6 times the amount of carbon dioxide we have today. Scientists believe this global warming was caused by continental collisions, events that released large amounts of this greenhouse gas into the atmosphere. (This also shows how the rock cycle and tectonic processes can affect atmospheric conditions.)

What is the **Kyoto Protocol**?

An international agreement initiated by the United Nations, the Kyoto Protocol is a document that aims to reduce global emissions worldwide in an effort to stem and eventually reverse global warming. Adopted on December 11, 1997, the agreement has been signed by representatives from 184 nations. A central aspect of the Kyoto Protocol is establishing scheduled reductions in emissions, reducing greenhouse gases relative to 1990 levels by five percent over the course of the years 2008 through 2012. Countries participating in the agreement can earn emissions-reducing credits by either reducing pollutants produced in their own countries, or through a "carbon market," which is a system of emissions trading. In other words, a country that wishes to allow more emissions within their borders can purchase emissions credits from countries with lower levels and still be in compliance with the treaty. Countries can also earn credits by sponsoring emissions-reducing programs in foreign nations or building clean factories and power plants in other countries.

Why did the **United States refuse** to sign the **Kyoto Protocol**?

The United States refused to sign the Kyoto agreement in 1997, and again in 2001, for a number of reasons. One major reason was that China and India were exempt from the stringent pollution controls that would have been imposed on the United States, so U.S. representatives said this would allow for an unfair trade advantage. Furthermore, the U.S. government was convinced that the protocol would severely damage the economy in general, costing many Americans their jobs and making the nation more dependent on foreign energy suppliers.

How does the **United States** compare with **China and Russia** on **energy use and production**?

As of 2007, the United States gets 50 percent of its energy from coal and, in 2006, was burning three times as much oil (21 million barrels) a day as China. America burns slightly less natural gas annually than Russia, which uses 604 billion cubic meters per year.

What could happen if **permafrost melts** permanently because of global warming?

The problem with global warming is that, as the process continues, it tends to (to use a rather ironic term) snowball and accelerate. A big reason for this was recently discovered as a result of permafrost and peat bog surveys being conducted in places like Scandinavia and Alaska. Here, land that was once covered in permafrost is thawing out and becoming wetlands. This might sound very nice, except that, stored beneath this layer of permafrost, is a buildup of centuries of methane gas created by decaying peat and other plant matter. Once this thaws, the methane will all be released into the atmosphere, increasing levels of methane by as much as 50 percent by some estimates.

What **countries are threatened** the most from **rising sea levels**, and may cease to exist in the twenty-first century?

Low-lying island nations of the Pacific and Indian Oceans are most at risk, most notably Tuvalu and the Maldives. Other candidates for severe flooding and reclamation of coastal land by the impending sea include Bangladesh, India, Thailand, Vietnam, Indonesia, and China, affecting the lives of hundreds of millions of people.

Is global warming causing an **increase in severe hurricanes**?

A 2005 Massachusetts Institute of Technology study estimated that, because global warming is increasing ocean temperatures, hurricane frequency and intensity has increased by about 45 percent from just a few decades ago. In the 1970s, for example, the average hurricane season saw 10 hurricanes develop in the Caribbean area; by the 1990s, that number had increased to 18 annually. Hugh Willoughby, a researcher for Florida International University, has also asserted that hurricanes are getting stronger and more frequent; this has also become the position of scientists at the National Oceanic and Atmospheric Administration. Among the exceptions to

What unexpected cooling phenomena are happening because of global warming?

It might be counterintuitive, but as the globe seems to warm, temperature readings indicate summer maximum temperatures are on the decline. However, at the same time, winter minimum temperatures are on the rise, so that the overall average has been an increase in temperature. Meanwhile, up in the stratosphere, the temperature has been cooling even as surface temperatures warm up. The reason for this is the destruction of ozone gases, which absorb ultraviolet radiation, which generates heat. The less ozone, the more UV radiation gets through the stratosphere, so it does not have as much of a chance to warm up. Records taken since 1979 show that three of the coldest stratospheric readings have been in 1997, 2000, and 2006.

this viewpoint is Florida State University professor James O'Brien, who has asserted that data demonstrates no significant increase trend between 1850 and 2005.

What **changes in plant and bird behavior** are indicating global warming?

People are noticing that spring is coming earlier, flowers bloom sooner, and summer has been lasting longer than in previous years, sometimes by a difference of one or more weeks. Bird migrations north and south have also changed, with bird watchers seeing species increasingly in northern climates where they have not before ventured. In North America, mammals such as opossums and armadillos are also increasing their northern ranges, and—a bit more worrisome—insects such as fire ants and Africanized bees are also moving north.

How much **land disappears** when the **sea level rises**?

Scientists believe that as the sea water rises by 0.04 inches (1 millimeter), the shoreline disappears by 4.9 feet (1.5 meters). This means that if the sea level rises by 3.28 feet (1 meter), the shorelines will extend another 1 mile (1.6 kilometers) inland.

Will **water supplies be threatened** by global warming?

There are a number of regions around the globe where populations get their drinking water from snow melting off of nearby mountains and draining into rivers. One area in the United States particularly threatened by global warming is Southern California. The snow on top of the nearby Sierra Nevadas is disappearing, and winters are not bringing the snow back as much as they once did. Los Angeles and other cities are increasingly plagued by water shortages as a result. Current estimates guess that 25 percent of the snowpack in the state will be gone by 2050, causing major stress on a state that has seen massive population growth. Meanwhile, drought in the Southwest has caused lakes and rivers to dry up. Economic losses as a result

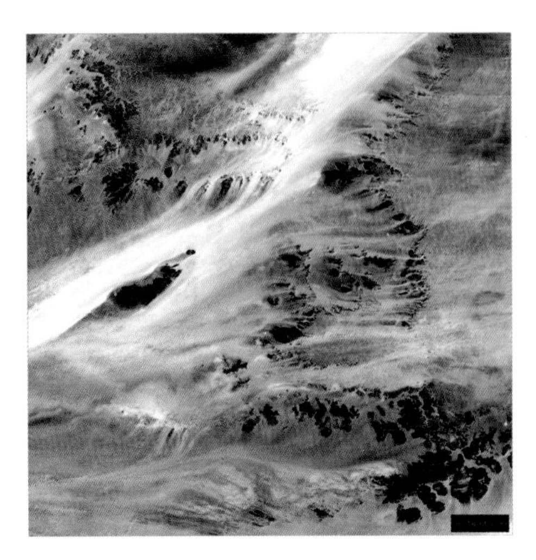

The Sahara Desert, just 10,000 years ago, was a rich grassland habitat populated by elephants, lions, and many other species. Today, it is the world's largest desert, and it is expanding at an alarming rate. (*NASA*).

of crop damage, livestock deaths, and loss of tourism in the region were estimated to be around $1.3 billion.

Is **global warming** causing **deserts to expand**?

Desertification—the expansion of deserts and corresponding decrease in fresh water supplies—is the result of several factors, including agricultural misuse, overpopulation, drought, and also climate change. There is no question whatsoever that the world's deserts are increasing in acreage at an alarming rate and that freshwater lakes and other water supplies are shrinking and turning to dust. Many examples may be cited, but here are just a few:

- Lake Chad, which has water shared by the African nations of Nigeria, Niger, Cameroon, and Chad, is 95 percent smaller than it was in the 1960s.
- Kazakhstan has lost half of its productive farmland since 1980.
- Sandstorms in Iran have engulfed more than 100 villages since 2002.
- The Gobi Desert in China is growing quickly, destroying land and causing giant dust storms that cost the country $6.5 billion every year.
- It is estimated that, by 2025, three fourths of Africa will face starvation because of the loss of water and farmland.
- Population migration and fights over resources are leading to wars, especially in Africa, such as the current conflict in Sudan.

What important **power source doesn't cause global warming**?

Although many people worry about what to do with the resulting waste products, nuclear energy has the advantage of not generating global-warming emissions. This fact has helped give a tremendous boost to U.S. nuclear reactors. But while the reactors themselves do not create emissions, such plants still leave a carbon footprint. This is because the uranium used to power them needs to be mined, the ore has to be shipped and processed, and plants have to be built and maintained. All of these things result in energy being used from sources such as gasoline and coal. As of 2008, there are 104 nuclear power plants in the United States providing a little over 20 percent of the nation's electricity needs.

Are **wind farms** a good way to solve the **energy crisis** and **global warming**?

Power from wind farms seems like it would be a very good solution to at least some of our energy needs. Wind, though unreliably fickle, is free, does not require min-

Many people think that wind farms are one solution to the energy crisis and global warming. However, even wind energy can have a negative impact on the environment and wildlife.

ing, and is clean. Also, windmills and generators have been improving to make this technology more economical. Currently, wind power supplies about 0.1 percent of the world's electricity, but that is growing at a rate of about 30 percent a year.

There are problems with wind farms, however. For one thing, the giant windmills you find in places like southern California kill birds and other animals. Any hapless bird who flies near one is apt to be chopped into bits. Some conservationists are worried this could actually kill some birds on the endangered species list. A 2002 study conducted in Spain of its wind farms noted that 350,000 bats, 3,000,000 small birds, and 11,200 birds of prey were killed by windmills and associated power lines *in just one year.* The Altamont Pass Wind Resource Area near Palm Springs, California, which consists of 4,900 windmills, is responsible for the deaths of 4,700 birds annually (about one per windmill), including 1,300 raptors (birds of prey) such as burrowing owls and golden eagles. Because of increasingly limited land space, some countries are building wind farms offshore. For example, England is constructing a new wind farm that will cover 145 square miles (375 square kilometers) off the Kent, Essex, Clacton, and Margate coastlines. Such offshore farms, of course, now pose a threat to seabirds.

Another fact is that wind farms take up a lot of space. Situated on hillsides and open plains, they effectively destroy what could be wildlife habitat. Each windmill takes up about 1,600 square feet (144 square meters) of space. Furthermore, building the vast wind farms needed to power cities takes a lot of resources, including the steel, concrete, and other materials needed to build and maintain windmills and the oil and gasoline that is burned during construction.

Do **wind farms** affect the **weather**?

Yes, to a degree. A 2008 study showed that large wind farms actually have an impact on local weather and climate. When one thinks about it, this makes sense. Farmers have for years been using large fans to decrease humidity and warm crops when there is a risk of a freeze. Wind farms, too, cause a decrease in humidity and raise temperatures, especially during the morning hours. The question becomes, as we construct more and more wind farms, how many windmills will it take to make a worldwide impact on weather and climate?

Is **global warming irreversible**?

A report released in January 2009 by the National Oceanic and Atmospheric Administration asserted that, even if we stopped carbon dioxide and other emissions completely this year, it is too late to go back. So many global warming gases have accumulated in the atmosphere that we are now destined to experience a period of warming that will likely last over the next 1,000 years.

Do **all scientists agree** that global warming is caused by human beings?

In short, no, not every scientist believes people create global warming. Although more and more scientists are becoming convinced that we are seeing a significant change in our climate, there are those who argue that it has more to do with solar cycles. Some meteorologists have also noted that, historically, climate warming has *preceded* increases in carbon dioxide levels, not the other way around. This is because when the Earth warms up, it also warms the oceans, which hold much of the world's carbon dioxide in the form of carbonic acid. When the oceans get warm enough, the carbon dioxide is then released into the atmosphere.

Hey, wait a minute, weren't scientists warning us of an **impending ice age** back in the **1970s**?

Beginning in 1970, there was considerable media attention given to speculation that Earth would soon see a new ice age. A scientific paper published in 1971 by S.I. Rasool and S.H. Schneider at the Institute for Space Studies is often cited as predicting an ice age. The paper was about the effects of aerosol levels in the atmosphere; the authors speculated such levels would rise by 600 to 800 percent over the next few years, and that this would trigger an ice age. What actually happened was that aerosol levels fell. Even if they hadn't, many scientists believe that Rasool and Schneider were inaccurate in their estimates about how carbon dioxide levels affected temperatures. Nevertheless, this publication and others that cited it drew the attention of the media, leading many people to believe that the next big climactic change would be a cooling off period.

Do other **scientists** believe we may be heading for an **ice age**?

Yes. In 2006, for example, chief researcher Khabibullo Abdusamatov at the Russian Academy of Sciences said that he believed worldwide average temperatures would

Could a sudden ice age like the one in the 2004 movie *The Day after Tomorrow* actually result from global warming?

No. That's Hollywood. In the movie, a scientist predicts (correctly for the plot of the film, which provides a lot of opportunities for special effects) that all the fresh water melting into the oceans will cause the Gulf Stream to dissipate, and that this will lead to the sudden onset of a new ice age. New York City freezes over in a day, and bizarre storm cells cause people to turn into popsicles in seconds. Ridiculous, say experts, though they admit that ice caps melting will indeed cause many adverse effects, such as rising oceans and changes in weather patterns.

That said, it is interesting to note that ice ages *can* occur fairly quickly. Geologists and glaciologists estimate that some ice ages made significant advances through North America in spans ranging from a few decades to as little as three years.

slowly decrease from 2012 to 2015, and then in the middle of this century there would be a much more precipitous drop in temperatures that would last for 60 years.

I heard that **some glaciers in Alaska are growing**, not shrinking. Is this true?

It's true that many Alaskan glaciers, including, most notably, the Hubbard Glacier, have been advancing in recent years. The reasons for this are complex and may have nothing to do with whether or not climate change is occurring. To simplify what is the rather complicated science of glaciology, there are different types of glaciers. Some glaciers rest in valleys and tend to be more sensitive to changing temperatures. Others, including Hubbard, are known as calving glaciers. They terminate at an ocean and parts of them break away in a stunning spectacle known as "calving." The five large glaciers in Alaska that are growing, including Hubbard, all have several commonalities: 1) they have previously experienced long periods of retreat only recently reversed (in the past century or so); 2) they all calve on shallow moraine shoals; 3) they all lay at the heads of long fjords; 4) they have positive mass balances that, by sheer force of weight, cause them to expand as gravity pushes down on them; and 5) they have small ablation areas; that is, small surface areas where melting and sublimation occur. Glaciologists note that glaciers with these qualities are not strongly influenced by even long-term temperature changes.

What **petition** have **thousands of scientists signed** that says that global warming is not caused by people?

Over 31,000 American scientists have signed the Global Warming Petition, which states that global warming is not caused by human activities, that there are actually many benefits to having a warmer planet, and that the United States should con-

Glacier calving in Glacier Bay, Alaska, is a natural process; however, it seems to be accelerating as glaciers melt faster because of global warming. (*photo by John Bortniak, NOAA Corps*)

tinue to reject the Kyoto Protocol. Looking at trends in such areas as ocean temperatures, solar activity, glacier shrinkage, and severe storm events, these scientists assert that upwards trends in all of these areas have been occurring since the 1800s. They believe, therefore, that policies limiting industrial and economic development are misplaced.

What was the **weather like** in **China in 2008**?

The winter of 2008 was one of the coldest on record in China. Even cities in South China, including Hong Kong, experienced record lows and power outages that sometimes lasted weeks. Over 100 people died as a result of the frigid weather.

Wasn't **2008** one of the **coolest years** for the United States?

Yes, for the United States the 2008 temperature averages were the lowest since 1997. Also, precipitation was way up that year, which, along with it being a La Niña year, is a primary reason why meteorologists believe it was a considerably cooler season. The average temperature from January to October 2008 was 55.9°F (13.3°C) versus 55.7°F (13.2°C) in 1997. This made 2008 actually a fairly average year with regard to records kept over the last 114 years. However, global temperatures were the ninth highest on record in 2008.

What has the **National Climatic Data Center** reported about average temperatures in **2008**?

According to the NCDC, the average temperatures for the contiguous 48 U.S. states was 0.3°F (0.14°C) cooler than the average measured from 1900 to 2000.

FORECASTING

When did **modern weather forecasting** begin?

On May 14, 1692, a weekly newspaper, *A Collection for the Improvement of Husbandry and Trade,* gave a seven-day table with pressure and wind readings for the comparable dates of the previous year. Readers were expected to make up their own forecasts from the data. Other journals soon followed with their own weather features. In 1771, a new journal called the *Monthly Weather Paper* was completely devoted to weather prediction. In 1861, the British Meteorological Office began issuing daily weather forecasts. The first broadcast of weather forecasts was done by the University of Wisconsin's station 9XM at Madison, Wisconsin, on January 3, 1921.

When was the first **official storm warning** issued in the United States?

Professor Increase Lapham, working with the U.S. Signal Corps, issued the first severe storm warning in America on November 8, 1870. The warning concerned a storm that was building strength in the Great Lakes.

What are **cooperative weather observers**?

While the U.S. government funds many weather stations throughout the country, the cost of supporting all of the stations that would be necessary to fully observe the weather *everywhere* would be prohibitive. Thankfully, volunteers known as cooperative weather observers assist by taking readings and measurements about winds, temperatures, precipitation, and so forth, providing this data to meteorologists in cooperation with the National Weather Service and the National Climatic Data Center.

Who came up with the **idea for volunteer weather observers**?

That credit goes to American physicist and mathematician Joseph Henry (1797–1878), the second president of the National Academy of Sciences who was also the first secretary of the Smithsonian Institution. Henry made advances in the area of electromagnetics, which led to his study in electromagnetic relays, which, finally, was the basis for Samuel Morse's (1791–1872) invention of the telegraph. While secretary at the Smithsonian, it dawned on Henry that the wonderful new telegraph could be used to link together weather observers throughout the country who could then relay the information to Washington, D.C. This became the network of volunteer observers that we have today.

How do meteorologists **predict weather** using **changes in air pressure**?

Meteorologists can forecast many details about the weather based on barometric readings. Generally, changes in air pressure indicate the following:

Decreasing pressure foretells rainy, windy, stormy weather.

Small, quick drops in pressure are good indicators that short periods of wind and rain will follow.

Slow, moderate drops in pressure indicate a low pressure system is in the area but this will not likely cause severe weather.

Slowly decreasing pressure over a longer period of time foreshadows poor weather that will last for some time.

Slowly decreasing pressure that was preceded by high pressure means that poor weather will be particularly bad.

Increasing pressure is an indication of dry, colder weather.

Slow, large rises in pressure mean that good weather is approaching and will likely last a long time.

Rapidly rising pressure when the pressure is already low is a good indicator of upcoming fair weather.

Rapidly decreasing pressure is a good prediction that a storm will hit within six hours.

Is there a difference between ***The Farmer's Almanac*** and ***The Old Farmer's Almanac***?

The Old Farmer's Almanac was first published in 1792 under the editorial leadership of Robert B. Thomas (1766–1846) and is currently published out of Dublin, New Hampshire. The similarly titled *Farmer's Almanac* debuted in 1818 in Ohio; the founding editor was David Young (d. 1852), and the publication is now headquartered in Lewiston, Maine. They are, indeed, different publications, although more confusion arises from the fact that, initially, *The Old Farmer's Almanac* was called *The Farmer's Almanac* for a number of years. Both, though, are almanacs, which means they include information about upcoming astronomical events, such as tides, sunrises and sunsets, lunar cycles, and so on. For added interest, they include cook-

Can groundhogs accurately predict the weather?

Over a 60-year period, groundhogs have accurately predicted the weather (i.e., when spring will start) only 28 percent of the time on Groundhog Day, February 2. Groundhog Day was first celebrated in Germany, where farmers would watch for a badger to emerge from winter hibernation. If the day was sunny, the sleepy badger would be frightened by his shadow and duck back for another six weeks' nap; if it was cloudy he would stay out, knowing that spring had arrived. German farmers who emigrated to Pennsylvania brought the celebration to America. Finding no badgers in Pennsylvania, they chose the groundhog as a substitute.

ing recipes, gardening tips, nature news, and advice columns. Published annually, they also both make predictions about the weather for the coming year, which is what made them desirable books for farmers to own. Both books claim to have secret formulas for predicting the weather the most accurately, with *The Old Farmer's Almanac* claiming it has an accuracy of 80 percent. Meteorologists dispute such claims, though, and also note that the almanacs make broad generalities about weather forecasts that makes them hard to refute. For instance, a prediction for spring might say that the Midwest will experience more rain than usual.

Can weather be predicted from the stripes on a **wooly-bear caterpillar**?

It is an old superstition that the severity of the coming winter can be predicted by the width of the brown bands or stripes around the wooly-bear caterpillar in the autumn. If the brown bands are wide, says the superstition, the winter will be mild, but if the brown bands are narrow, a rough winter is foretold. Studies at the American Museum of Natural History in New York failed to show any connection between the weather and the caterpillar's stripes. This belief is only a superstition; it has no basis in scientific fact.

Are there **trees** that **predict the weather** and tell time?

Observing the leaves of a tree may be an old-fashioned method of predicting the weather, but farmers have noted that when maple leaves curl and turn bottom up in a blowing wind, rain is sure to follow. Woodsmen claim they can tell how rough a winter is going to be by the density of lichens on a nut tree. Before the katydid awakes, a black gum tree is able to indicate the oncoming winter. Trees can also be extraordinary timekeepers: Griffonia, in tropical west Africa, has two-inch (five-centimeter) inflated pods that burst with a hearty noise, indicating that it is time for farmers of the Accra Plains to plant crops; Trichilia is a 60-foot (18-meter) tree that flowers in February and again in August, signaling that it is time, just before the second rains arrive, for the second planting of corn. In the Fiji Islands, planting yams is cued by the flowering of the coral tree.

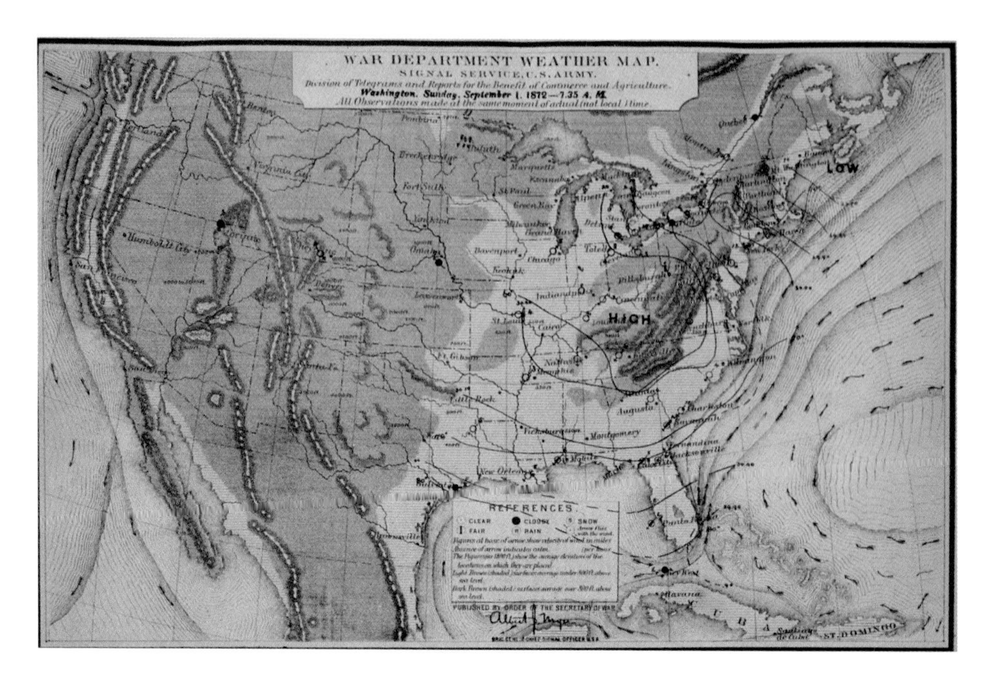

This weather map, issued on September 1, 1872, was produced by the U.S. Army Signal Service, and displays information such as air pressure, cloud cover, precipitation, and ocean currents, but only for the eastern half of the United States. (*NOAA*)

Is a **halo** around the **Sun or Moon** a sign of rain or snow approaching?

The presence of a ring around the Sun or, more commonly, the Moon in the night sky, betrays very high ice crystals composing cirrostratus clouds. The brighter the ring, the greater the odds of precipitation and the sooner it may be expected. Rain or snow will not always fall, but two times out of three, precipitation will start to fall within 12 to 18 hours. These cirroform clouds are a forerunner of an approaching warm front and an associated low pressure system.

When did the **first weather map** appear in a **newspaper**?

The London *Times* was the first periodical to publish a weather map, which ran in the April 1, 1875, issue. The map was created by Sir Francis Galton (1822–1911), an English scientist and half cousin of Charles Darwin. Galton created his map—which showed prevailing winds, barometric pressures, and temperatures throughout the British Isles and parts of Western Europe—without the advantage of satellites, computers, or even a telephone.

When did **maps** displaying **cold and warm fronts** first get published in the United States?

Although the discovery of weather fronts was made during World War I, it was not until the 1930s that U.S. weather maps began to use symbols to indicate them.

What is "nowcasting"?

Nowcasting is a somewhat silly-sounding word that refers to predicting weather in the very short term: about two hours. Given the satellites, radar, and other modern tools that are at meteorologists' disposal, nowcasting is probably the most accurate type of forecasting. Weather patterns, though they can still change abruptly with little notice, such as in the case of tornadoes, are fairly predictable in the short term. For example, if you see a large, well-organized storm front heading toward your city just a few miles away, it is a safe bet that a weather forecaster can tell you precisely when it will affect your neighborhood and what weather conditions you can expect.

What is the difference between a National Weather Service **advisory, statement, watch,** and **warning**?

The National Weather Service will issue a *statement* as a "first alert" of a major change in the weather. An *advisory* is issued when weather conditions are not life threatening, but individuals need to be alert to weather conditions. A weather *watch* is issued when conditions are more favorable than usual for dangerous weather conditions, e.g. tornadoes and violent thunderstorms. A watch is a recommendation for planning, preparation, and increased awareness (i.e., to be alert for changing weather, listen for further information, and think about what to do if the danger materializes). A *warning* is issued when a particular weather hazard is either imminent or has been reported. A warning indicates the need to take action to protect life and property. The type of hazard is reflected in the type of warning (e.g., tornado warning, blizzard warning).

What is the **Storm Prediction Center** (SPC)?

A part of the National Weather Service and the National Centers for Environmental Prediction, the Storm Prediction Center focuses its efforts on forecasting only hazardous weather, including heavy rains and snows, as well as conditions that could lead to dangerous wildfires.

How do **meteorologists** come up with **percentages predicting the weather**?

We've all heard weather broadcasts where a meteorologist will say something such as "This afternoon there will be a 40 percent chance of snow showers" or "A 75 percent chance of rain will likely make it a wet night," but how do they come up with these figures? To many audiences this might seem like mere guesswork on the part of the television weatherpeople. Actually, they arrive at such figures using a series of computer models. What typically happens is that a meteorologist will gather together as much data as possible from local and national weather stations, including from weather satellites, Doppler radar, temperature readings, etc., feed this data into a computer weather modeling program, and run dozens of scenarios of what could happen given these initial conditions. If, for instance, 20 out of 60 of the scenarios that are run predict that there will be rainfall, then the meteorologist will predict a 30 percent chance of rain. While chaos theory dictates that there is no reasonable way one can expect meteorologists to have a 100 percent—or even a 90 or 80 percent—accuracy rate, they come as close as possible given our current knowledge of how weather works and the limitations even computer technology has.

What do meteorologists mean when they talk about **POP**?

POP is an abbreviation for "probability of precipitation."

Is there a **difference** between a forecast of **partly sunny** or **partly cloudy** skies?

Yes. The difference is that "partly cloudy" refers to a sky that is mostly clear with some clouds in it, while "partly sunny" is a sky that has more cloud cover than clear patches.

What is **NOAA Weather Radio**?

The National Oceanic and Atmospheric Administration's Weather Radio is a 24-hour-a-day broadcast of weather conditions throughout the United States (about 90 percent of the lower 48 states are covered), including its territories. Working in conjunction with the Emergency Alert System and the Federal Communications Commission, as well as other federal, state, and local officials, Weather Radio also broadcasts warnings about other hazards, such as earthquakes and tsunamis, or environmental accidents (oil or chemical spills, for example), and public service announcements like AMBER Alerts about missing children.

What is **aviation forecasting**?

Aviation forecasting is a critical support service for the airline industry. Aviation forecasters warn pilots and airline management about potentially hazardous weather conditions that could lead to wind shear, wing icing, severe turbulence, strong winds, thunderstorms, or other dangers. While passengers might be frustrated by the delays in air travel that result from these warnings, aviation forecasting probably saves hundreds, even thousands, of lives every year. In addition to the worst-case scenarios that might occur if pilots were not warned of dangerous weather, aviation forecasters can also advise on wind conditions that can help save fuel by not having planes fly into head-on winds. When the cost of airline fuel rises, this has the potential of saving airlines billions of dollars.

What is **marine forecasting**?

The complement to aviation forecasting, marine forecasting warns sea-going vessels of storms and wave conditions. Such forecasting routinely saves lives, property, and fuel. The National Oceanic and Atmospheric Administration's National Data

Buoy Center keeps track of ocean conditions, particularly issuing hurricane warnings and warnings about swells that may indicate tsunamis caused by underwater earthquakes.

What is **agricultural forecasting**?

Agricultural forecasting concerns itself with predicting weather conditions, especially hail, that can be hazardous to crops, including precipitation (especially hail), extreme hot and cold, and damaging winds. Accurate forecasting in this area is a particular boon to farmers in that meteorologists can help them decide when it is best to harvest or plant, when wind conditions are right to apply pesticides and herbicides, when to set up irrigation systems, or when to turn on wind machines and smudge pots before an oncoming freeze.

What is **industrial forecasting**?

Industrial forecasting has applications in economics, both local and national. Weather affects businesses, transportation, and consumer activity in a wide variety of ways. Predicting heat or cold waves, for instance, can help utility companies plan for surges in consumers' use of air conditioners and furnaces. Cities located in drier climates, such as Los Angeles, benefit from rain predictions that affect water reservoirs; local governments, for instance, can issue water conservation guidelines in times of drought to prevent resources from being stretched too thin. The sports industry, a multibillion dollar market, benefits from forecasts planning for rain, snow, or other harsh weather conditions. Forecasts of ice and snow can help transportation firms plan shipments; and even the fast food industry is affected by weather. For example, research has shown that people order a lot more pizza deliveries in cold and inclement weather. In short, industrial forecasting is of great interest to business and government, as it helps people plan for potential losses (winter blizzards or hurricanes routinely cost local economies billions of dollars) or redistribution of resources.

What is **fire weather forecasting**?

Fire weather forecasters concern themselves with studying rainfall, humidity, temperature, thunderstorms, wind and sunlight conditions that could leave areas such as forests and grasslands vulnerable to wildfires. These forecasts can prepare fire crews and other emergency support before a fire begins, and after fires begin they can help professionals determine such things as the possible direction a fire will spread and whether or not an oncoming rain storm might help put the fire out.

What is **transportation forecasting**?

Most people are probably familiar with this specialty in forecasting, which helps warn car drivers and truckers that road conditions could be hazardous. It also helps prepare local governments to clear streets and freeways, spray road salt or sand, and have law enforcement and emergency personnel ready for emergencies. On the

A Coast Guard aircraft drops a hurricane warning announcement down to a sponge fishing boat off the coast of Florida in 1938. (*NOAA*)

business level, weather forecasts can warn shipping companies, for instance, not to ship perishable products in unrefrigerated or unheated trucks. Shipments by rail also need to be ready for unfavorable weather conditions. Some larger companies rely on private forecasting firms to provide them with timely advisories.

Can meteorologists **predict tornadoes**?

It is difficult to nearly impossible to predict where and whether or not a tornado will strike. All forecasters can do is warn people when conditions are right for tornadoes to form, or, if one is sighted, to tell people to take shelter. Meteorologists look for tornadic thunderstorms that have strong indications of wind shear, lift, moisture, and instability. No one type of weather pattern leads to tornado formation, which greatly complicates forecasting efforts. To aid in their predictions, meteorologists use all sorts of technology, including weather balloons, Doppler radar, satellites, data from weather stations, lightning strike plots, and computer modeling.

Can you **see a tornado** using **Doppler radar**?

No. Doppler radar can tell meteorologists if conditions within a storm are favorable for tornadoes—such as strong winds and cloud rotation—but it can't actually see a tornado.

Who were the **first people** to successfully **forecast a tornado**?

U.S. Army officers Ernest Fawbush and Robert Miller were the first to correctly predict a tornado would form on March 25, 1948. Recognizing that weather patterns in central Oklahoma were very similar to those that had occurred a few days earlier when a tornado hit Tinker Air Force Base, Fawbush and Miller told their superiors and a decision was made to warn residents about the possible threat. A tornado again hit the Tinker base a few hours later.

What is **numerical weather prediction**?

Numerical weather prediction—or numerical forecasting—is the science that believes that weather forecasting is possible if one has a thorough knowledge of the

Why was tornado forecasting once banned in the United States?

Concerns that warning people about possible tornado formation would panic residents led the Weather Bureau to ban such forecasting off and on in the 1940s. As meteorology improved and tornadoes became somewhat less terrifying in their ability to appear suddenly, the ban was lifted in 1950.

laws of physics and also knows the current state of the weather. Proposed by a group of Norwegian scientists collectively known as the Bergen School, the idea was that air behaves much like a fluid, and that it therefore adheres to the hydrodynamical equations that liquids like water do. Knowing the current state of the weather is vital, and so numerical forecasting relies heavily on having detailed weather reports from multiple locations before predictions can be made. Once this is available, mathematical formulas are applied to the weather's current state based on the principles of thermodynamics, the Boyle's law, Newtonian physics, and so on.

Who **first proposed** the approach of **numerical weather prediction**?

Vilhelm Bjerknes (1862–1951), a Norwegian physicist and meteorologist, was the author of the first formal studies on weather forecasting. He was also famous for his seminal 1921 book, *On the Dynamics of the Circular Vortex with Applications to the Atmosphere and to Atmospheric Vortex and Wave Motion,* which proposed the fundamental ideas behind numerical weather prediction in 1904. The theory was picked up again in 1922 by English mathematician and meteorologist Lewis F. Richardson (1881–1953). The mathematics behind Bjerknes's theory appealed to Richardson, but the calculations necessary to come up with the predictions, in a time before the invention of the computer, were formidable. Richardson estimated it would take a coordinated effort of about 26,000 people using calculators to figure out the math fast enough for the numerical method to work. Trying to do some preliminary calculations himself, Richardson's early attempts at weather predictions were far off the mark. His misunderstanding of some of Bjerknes's numerical methods led him to come up with estimates on air pressure that were far too high. Because of Richardson's failure, numerical weather prediction was abandoned until the 1940s.

What happened in the **1940s and 1950s** that gave hope to the science of **numerical weather prediction**?

Hungarian-American mathematician John von Neumann (1903–1957) devised a forerunner of the modern computer that could make the rapid calculations needed to predict the weather using the numerical forecasting method. Next, Princeton University meteorologist Jule Charney (1917–1981), having studied Richardson's earlier failure, wrote revised formulas in 1946 that could be used for weather prediction with the help of von Neumann's computer. With this background foundation in place, in 1950 the first successful weather forecast using the numerical method was completed in April, 1950, by the ENIAC computer at Maryland's Aberdeen Proving Ground. An ongoing weather forecast service was then begun in 1955, using an IBM computer funded by the National Weather Service, and the U.S. Navy and Air Force.

RADAR

What is **radar**?

Radar is an acronym for "RAdio Detection And Ranging," and it uses radio waves to detect objects in the atmosphere. It was first devised in 1904 by the German inventor Christian Hülsmeyer (1881–1957), who called his radar detector a "telemobiloscope" and patented the device in 1906. The original purpose of his invention was for ships to be able to detect each other so that in poor weather conditions (e.g., heavy fog) they would not run into each other. Sadly, this brilliant invention did not catch on at the time; if it had, some speculate that the 1912 *Titanic* disaster could have been avoided. Another concept that Hülsmeyer came up with was the remote control. He believed, correctly, that radio waves could be used to turn mechanical devices on and off. Again, people ignored the concept and he never got the credit he deserved.

When did **radar** finally catch on, and who **gets the credit**?

In 1935—over three decades after Hülsmeyer first proposed the idea—radar technology came into its own through the work of British scientist Robert Watson-Watt (1892–1973), along with H.E. Wimperis (1876–1960), Henry Tizard (1885–1959), and A.F. Wilkins. Working for the British government, Watson-Watt was given the assignment of investigating whether Germany's Adolf Hitler could carry out a threat of creating a weapon using radio waves. Watson-Watt knew this was an impossibility, but he saw another potential use. Using shortwave radio transmitters from the British Broadcasting Corporation, he and his colleagues created the first practical radar technology. Radar was used to detect attacking German airplanes during World War II, and was credited with swaying the 1940 Battle of Britain in England's favor.

Who should be **credited** with **inventing weather radar**?

No one person decided to use radar for weather forecasts. With the technology already in place, it was simply adapted to this purpose when experiments in Britain and the United States showed that radio waves bounced off clouds. Radar was first used to specifically obtain weather data in 1949, but it was not until the mid-1950s that a weather station using radar technology was established in the United States. This happened after the Eastern seaboard was hit by two vicious hurricanes in 1954 and 1955. The U.S. Weather Bureau was then authorized by Congress to create a national weather radar grid, and so the Weather Surveillance Radar (WSR-57) was founded in 1957. WSR systems used vacuum tubes and other technologies that were

The first National Severe Storms Laboratory radar, shown in this 1971 photo, was constructed in Norman, Oklahoma. This early radar, with a 30-foot (3-meter) dish, eventually led to the development of the NEXRAD WSR-88D radar. *(NOAA Photo Library, NOAA Central Library; OAR/ERL/National Severe Storms Laboratory)*

becoming outdated by the late 1970s. Despite the fact that vacuum tubes were in short supply, and other parts had to be hand-machined in order to keep the weather system running, Congress did not approve replacing the system until the 1990s.

What is **Doppler radar**?

Doppler radar measures frequency differences between signals bouncing off objects moving away from or toward it. By measuring the difference between the transmitted and received frequencies, Doppler radar calculates the speed of the air in which the rain, snow, ice crystals, and even insects are moving. It can then be used to predict speed and direction of wind and amount of precipitation associated with a storm.

What **replaced the WSR-57s**?

WSR-74s were added to the radar grid in 1974 and were used in conjunction with the older "57s," as they were called. It was not until 1988, however, that the Next Generation Radar (NEXRAD; also called WSR-88D) were built using the newer Doppler technology. The first NEXRAD was deployed in Norman, Oklahoma, in 1990. There are now 160 NEXRAD radar stations throughout the United States.

What is the **Doppler effect**?

The Doppler effect refers to the way sound and light waves behave, depending on the motion of the transmitting and receiving objects. As objects move away from each other, waves become stretched out; and as objects move toward each other, waves become compressed. When it comes to sound waves, the result is that one hears a higher-pitched sound if an object making noise comes closer; if that object is moving away, the result is a lower-pitched sound. When it comes to light waves, compressed visible light is shifted toward the blue spectrum, while it is red-shifted if objects are moving away. Astronomer Edwin Hubble (1889–1953) used this information about light waves to form his theory of an expanding universe.

The Doppler effect gets its name from Austrian physicist and mathematician Johann Christian Doppler (1803–1853), a figure in history whose real name has caused considerable confusion. His baptismal certificate indicates that he was actually born Christian Andreas Doppler, but on his gravesite his name is presented as Christian Johann Doppler. Depending on sources, he has also been cited as Christian Andreas Doppler and Johann Christian Andreas Doppler! Let's just go with his last name. Interestingly, the Doppler effect was not confirmed by its theorizer. Rather, it was a Dutch scientist named Christoph Hendrik Diederik Buys Ballot (1817–1890) who proved it by conducting an experiment in which he was trying to *refute* Doppler's theory. Buys Ballot set up a series of trials by a railroad track in Utrecht, the Netherlands. An accomplished horn player road a train while a second highly trained musician stood by the tracks, listening to the first musician play a note on the horn. After two trials called on weather, a third trial confirmed that Doppler was correct about how sound waves change with motion.

How do **radars** use the **Doppler effect** to measure the weather?

Doppler radar—unlike conventional radar systems that could only indicate where a storm was and how intense it was—can be used to detect wind speed and the direction storms are taking. Because one Doppler radar can only tell if a storm is heading away or toward it, a network of radars is needed to chart a storm's precise course.

What do the **colors** shown on **television radars** indicate?

Colors on weather maps shown on television news programs are indications of precipitation levels. Cooler colors (blue, green, etc.) indicate light precipitation, while warmer colors (yellow, orange, red) are used for heavier precipitation.

What is a **hook echo**?

A hook echo—also called a "hook pattern"—is a warning sign that a tornado has possibly formed. On radar, the echo looks kind of like the number "6" and is associated with a mesocyclone within a storm, so it is called a tornadic mesocyclone. The distinctive hook shape was first seen on April 9, 1953, by electrical engineer Donald Staggs, who was monitoring the radar system at the Illinois State Water Survey in Champaign. A hook echo does not necessarily indicate a tornado has definitely formed, but it is a strong indication that it has. On the other hand, tornadoes can and do form without the tell-tale hook echo ever being seen.

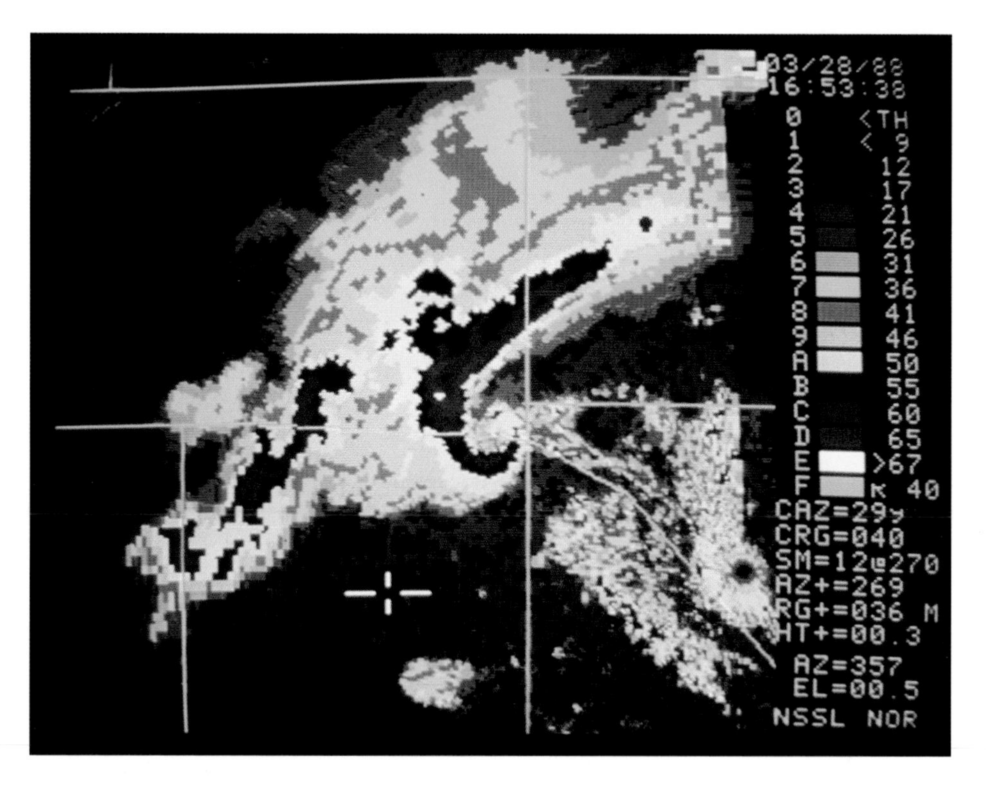

A 1988 radar image of clouds near Norman, Oklahoma, shows the distinctive hook echo that is often a precursor to tornadic activity. (*NOAA Photo Library, NOAA Central Library; OAR/ERL/National Severe Storms Laboratory*)

Has **Doppler radar** made **tornado forecasting** better?

Although predicting tornadoes is still far from a perfect science, Doppler radar used by the NEXRAD system has improved the accuracy of prediction to about 80 percent (versus about 30 percent before NEXRAD). Smaller, quickly forming tornadoes are still nearly impossible to anticipate, but larger tornadoes, especially huge tornadoes in the EF4 or EF5 category range, are being anticipated with much better results. In fact, the National Weather Service is often able to warn nearby residents of a large tornado 20 or more minutes before it actually strikes. NEXRAD can do this by searching for tornadic mesocyclones, which are a good indicator of twisters. Any large tornado vortexes can actually be seen by Doppler radar these days. Thanks to these advances, it is much less likely that a big tornado will take nearby populations completely by surprise. Even a 10- or 20-minute warning can be enough to save dozens or even hundreds of lives.

What is a **Pulse-Doppler radar**?

Pulse-Doppler radars—or simply "pulse radars"—are designed to better track the velocity of clouds, winds, and precipitation. They use radar in short, strong bursts with long intervals in between. To get accurate measurements, several pulse radars must be used in conjunction.

Scientists release a rawinsonde balloon, while weather instruments and a mobile radar unit (background) also monitor humidity, air pressure, precipitation, and temperature. (*NOAA Photo Library, NOAA Central Library; OAR/ERL/National Severe Storms Laboratory*)

What is **lidar**?

You can think of lidar as being the equivalent of a light radar. Meteorologists can use laser beams pointed at the atmosphere to measure such things as pollution concentrations and wind velocity.

What is **sodar**?

If lidar is light radar, then you can probably guess that sodar is sound radar. Sodar stands for Sound Detection and Ranging. It is different from sonar in that sonar uses sound waves traveling through water, whereas sodar is detecting the reflection of sound waves through the air. By calculating the Doppler shift of sound waves, sodar can be used for evaluating wind speeds and direction, as well as temperature inversions and other types of turbulence.

What is a **Profiler**?

Profilers use electromagnetic waves the same way that sodars use sound waves. This strategy has proven ideal for penetrating deep into the upper atmosphere, where meteorologists can then determine wind speeds at altitudes 5 to 10 miles (8 to 17 kilometers) high.

What is a **radiosonde**?

The idea of using balloons to aid in the study of the atmosphere was first explored by the French when, in 1784, a hot air balloon was used for this purpose. It took a long time, however, before the practice came into common use. More commonly thought of as a "weather balloon," radiosondes ("sonde" is French for "probe")

How were kites once used in early meteorology?

Kites, when skillfully designed and flown, can reach extraordinary heights. The world record for a single kite being flown is 14,509 feet (4,422 meters). The record was set in Kincardine, Ontario, Canada, on August 12, 2000, by Richard Synergy. Much greater heights may be achieved, however, when multiple kites are linked together. A chain of 10 kites set an American record in 1910, rising to 23,385 feet (7,128 meters), and a collection of eight kites reached 31,955 feet (9,740 meters) in Lindenberg, Germany, in 1919. The earliest record of kites being used for meteorology comes from the University of Glasgow, where in 1749 two students, Alexander Wilson and Thomas Melville, wanted to measure conditions in the upper atmosphere. Since there were no airplanes or hot air balloons at the time, they used a kite.

are collections of weather-detecting instruments attached to a balloon that is released into the upper atmosphere. They were first used in Europe during the 1920s and 1930s. Radiosondes typically measure temperature, moisture, and wind speeds, and they often include small, battery-powered motors. More modern radiosondes—called "rawinsondes"—include a radar reflector so they can be more easily tracked.

Radiosondes reach elevations that can take them into the stratosphere. Once it reaches its maximum height, the balloon will burst and the instrument package will be carried safely down on a parachute. Another way to deploy a radiosonde is by dropping it from an airplane. When this is done, the device is called a dropsonde. Rocketsondes—weather probes mounted on rockets, as one might guess—may also be used on occasion. There are over 800 radiosonde launch sites around the world, with the probes being launched at midnight and noon. Data is generally shared by meteorologists around the world.

SATELLITES

How were the first **high altitude photographs of clouds** taken?

During World War II, cameras were mounted aboard some of Germany's V2 rockets to take pictures of cloud patterns. The success of these efforts inspired meteorologists to plan for weather satellites.

What was the **first satellite** used for monitoring the weather?

The first man-made satellite used to monitor weather conditions was the Television and Infrared Observation Satellite (TIROS I), which was launched by NASA on April 1, 1960. While the photographs taken were not of the high-resolution standards we

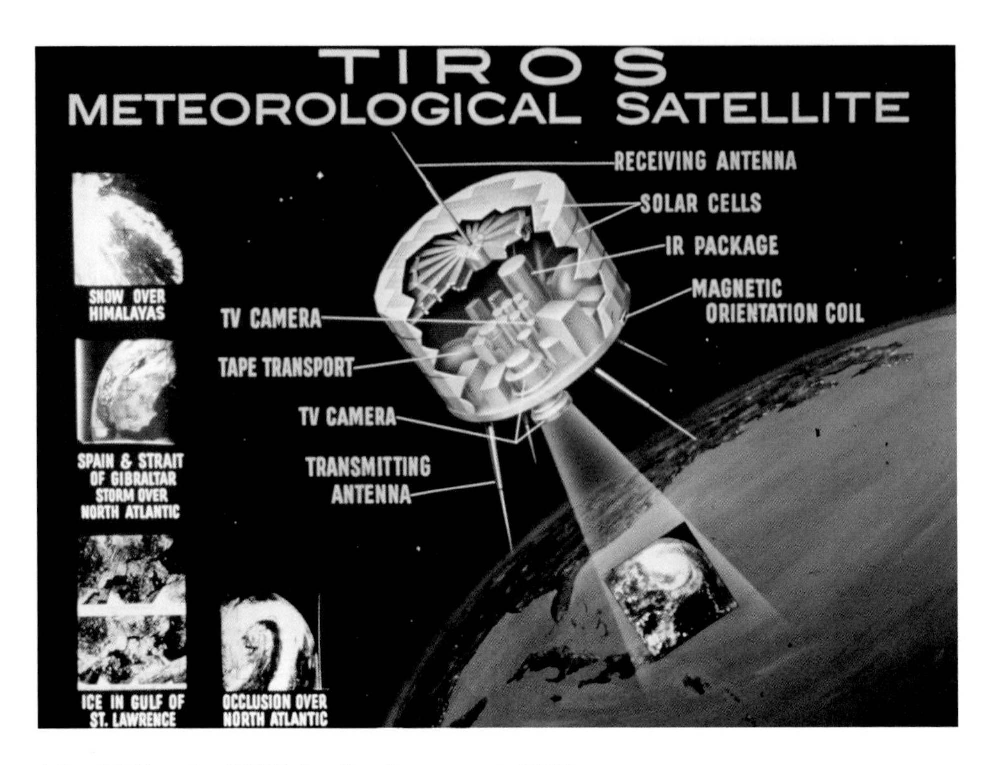

A circa 1961 illustration of TIROS I shows its various components. (*NOAA*).

see today, they were the first to reveal just how clouds and storms can be remarkably well organized, a fact that surprised meteorologists at the time. TIROS I's other groundbreaking accomplishment was to spot a previously undetected tropical storm near Australia nine days after its launch. Australians along that country's east coast were thus the first people, thanks to modern technology, to get a heads up that a strong storm was approaching.

What **remarkable observation** did science fiction author **Arthur C. Clarke** make?

Arthur C. Clarke (1917–2008) is well known to science fiction fans. His 1948 short story "The Sentinel" was the basis for the 1968 film *2001: A Space Odyssey*. Among his many accomplishments, he was also very interested in satellites. During World War II, Clarke was a radar technician for the Royal Air Force, and in 1945 he proposed designs for a communications system using satellites. He reasoned that this was possible if satellites could be placed in orbit above the equator while traveling 22,248 miles (35,797 kilometers) per hour. This would put them in geostationary orbit, meaning that each satellite would remain directly above a predetermined point on the Earth's surface. This idea proved correct, and is used now for both communications and weather satellites. The Clarke Belt, a band of space over the equator at an altitude of 22,300 miles (35,800 kilometers) where geostationary satellites may orbit, is named in his honor.

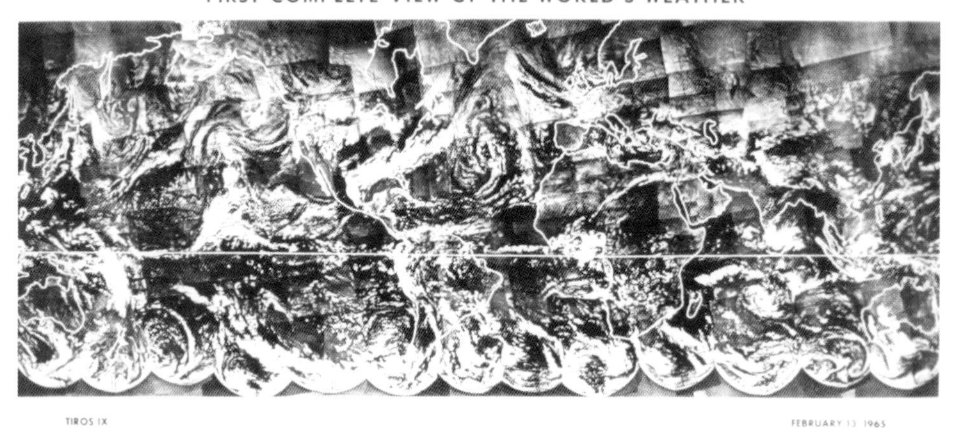

A composite of photos from the TIROS IX, this image from February 13, 1965, is the first complete view of our planet's weather. (*NOAA*)

What are the **different types of satellites** used today?

The two types of satellites used for weather observations are geostationary and polar orbiting. Geostationary satellites remain above a fixed point on the Earth's equator by traveling at a speed that matches that of the planet's spin; polar orbiting satellites circle the Earth from pole to pole. These satellites orbit at a speed so that, two times a day, they are over the exact same spot on the planet in what is called a "Sun-synchronous" orbit.

What is **POES**?

POES is an acronym for Polar Operational Environmental Satellite. These are the successors to the TIROS satellites, and they are thus called the Advanced Television Infrared Observation Satellites (ATN or TIROS-N). Like the GOES, there are two POES operated by the U.S. Department of Defense's Air Force Space and Missile System Center in what is called the Defense Meteorological Satellite Program. The POES orbit the Earth at elevations ranging from 515 to 540 miles (830 to 870 kilometers) in altitude, with one crossing the equator at 7:30 A.M. and 1:40 P.M.. Data is then transmitted to stations located at Wallops Island, Virginia, and in Fairbanks, Alaska.

What is a **GOES**?

GOES stands for Geostationary Operational Environmental Satellite. The first GOES was launched in 1975. Currently, the National Oceanic and Atmospheric Administration (NOAA) operates two such satellites that were launched by the National Aeronautics and Space Administration (NASA): GOES-10 and GOES-12. There is also a backup satellite, GOES-11, which can be activated in the event that one of the other satellites malfunctions. GOES-10 is also known as GOES-West because it monitors the sky over the Western Hemisphere; and GOES-12 (GOES-

What is the Solar and Heliospheric Observatory (SOHO)?

SOHO is a satellite paid for and launched through a cooperative effort between NASA and the European Space Agency. Launched on December 2, 1995, the satellite makes observations about the Sun's corona and solar winds, and it is also intended to try and study the Sun's inner core. SOHO has provided scientists with information about sunspot structure and solar wind speeds; the satellite has taken photographs of the Sun's turbulent convection zone, and it has discovered solar tornadoes and coronal waves.

East) monitors the Eastern Hemisphere. Each satellite is capable of viewing about one-third of the sky at any one time.

What **other countries** also have **weather satellites** in orbit?

Currently, the following countries have weather satellite programs: Japan, Russia, China, India, South Korea, and Europe (European Space Agency). Japan launched its Geosynchronous Meteorological Satellite *Himawari* in 1995, but in 2003 it malfunctioned, and so NOAA permitted the Japan Meteorological Agency use of the older GOES-9 satellite. The European Organisation for the Exploitation of Meteorological Satellites (EUMETSAT) operates what is now the second generation of METEOSAT satellites. Europe launched its first weather satellite in 1995, and this second generation came online in 2004 with METEOSAT-8. This satellite scans the entire globe every 15 minutes. China launched the *Feng-Yun* in 1990, and there have been several successors since then. That same year, Russia launched GOMS (Geosynchronous Operational Meteorological Satellite). The Indian Satellite (INSAT) made orbit in 1990, and was used for both weather observations and communications. It was followed by INSAT-2 through INSAT-4 series, before the KALPANA-1 was operational in 2002. The KALPANA-1 satellite is India's first exclusively meteorological satellite. South Korea's first weather satellite is the Communication, Ocean and Meteorological Satellite (COMS), which was launched in 2005.

How did the KALPANA-1 get its name?

India's KALPANA-1 satellite was originally called the METSAT (meteorological satellite). However, it was renamed in honor of Kalpana Chawla, the Indian astronaut who died in the space shuttle *Columbia* disaster on February 1, 2003.

What **measurements** do **weather satellites** provide us?

Early weather satellites transmitted images of clouds and could also take infrared (IR) readings, which allowed for weather monitoring both day and night. With IR sensors, satellites can detect temperature readings, which can also indicate cloud elevation. Modern satellites are capable of many other observations, as well. Detailed images of not only clouds, but also land and sea surfaces and temperatures, provide scientists with information about fog, snow, rain, ocean currents, haze, air pollution, ozone levels, soil moisture levels, airborne dust, and even volcano and forest fire activity. Satellites are also used to take surveys of agricultural activities and vegetation growth.

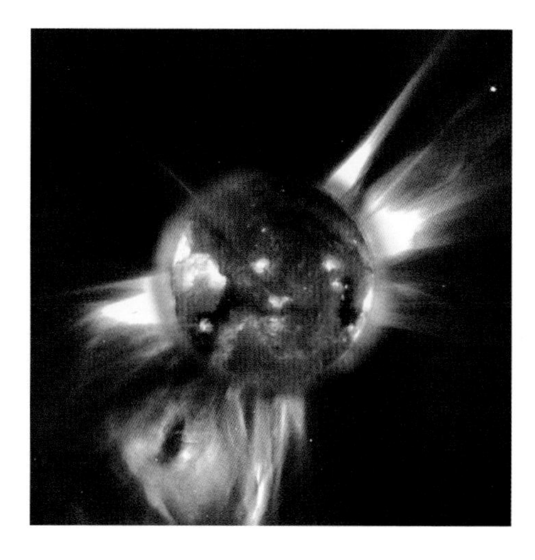

The SOHO spacecraft captured this image of the Sun shooting coronal mass ejections (CMEs) of plasma out into space. CMEs occur every week, and sometimes more than once a day. *(NASA/JPL)*

What is the **Landsat Program**?

In 1972, NASA launched the first Landsat (land satellite) in cooperation with the U.S. Geological Survey. While taking photos and measurements related to the planet's surface, these satellites also provide a considerable amount of weather-related information. For example, Landsat satellite data has recorded the impacts of weather disasters, monitors flooding, and takes other measurements of interest to hydrologists. Recent surveys of land use have also resulted in studies about the effects of human activities on weather. For example, in 2006, an atmospheric modeling study was conducted on the state of Florida. Scientists compared estimates of vegetation coverage in the year 1900 with present conditions and demonstrated how urban and rural activities were affecting rainfall. It was estimated that Florida now experiences 12 percent less rain than it would have a century ago.

CAREERS IN METEOROLOGY

What is a **meteorologist**?

People usually associate meteorologists with television station weathercasters. Most meteorologists, though, work behind the scenes. They can be found working for the National Weather Service, in laboratories, at weather research stations, or working for universities. Meteorologists need a good understanding of physics, chemistry, hydrology, and other disciplines to do their jobs well. The American Meteorological Society defines a meteorologist as someone who can research, observe, explain, and forecast the weather; who understands the principles behind weather phenomena; and who appreciates the effects that the weather has on Earth. Meteorologists obtain a bachelor's degree in science to do their jobs, and many of them have master's degrees or doctorates. There are also a variety of specialties in the field, such as hydrology and climatology, and many meteorologists study mathematics, computer engineering, electrical engineering, and more.

Do some people pursue meteorology as a hobby?

There are many weather enthusiasts in the United States and abroad. Meteorology, in fact, is one of the most popular avocations one can become involved with. From storm chasers to cooperative weather observers to ham radio operators, there is a large community of people who study the weather just for the fun of it. Two periodicals, *American Weather Observer* and *Weatherwise,* are popular resources for meteorology enthusiasts.

Is **meteorology** considered a **good career**?

Meteorology is certainly one of the better careers you can go into in terms of income, stress level, satisfaction, working environment, physical demands, and employment outlook. In its ranking of 250 careers in the United States, the sixth (2002) edition of the *Jobs Ranked Almanac* placed meteorologist as thirteenth. While this is somewhat lower than the seventh ranking it earned in the previous edition, it is a lot higher than the thirty-eighth place it had in the fourth edition, so it seems to be moving up.

What are the different **specialties within or related to meteorology**?

Meteorology isn't limited to weather prediction. There are numerous related areas that can combine weather with everything from engineering, forensics, and computer science to television media, storm chasing, and commodities trading. Here is a list of specialty fields that might be of interest to you.

- Acid precipitation researcher
- Agricultural forecaster
- Air quality forecaster
- Air quality modeler
- Air traffic control assistant
- Atmospheric chemist
- Atmospheric optics researcher
- Aviation forecaster
- Bioclimatologist
- Broadcaster
- Climatologist
- Commodities trader
- Computer visualization specialist
- Data communications engineer
- Educator

A scientist works in a National Severe Storms Laboratory mobile unit, processing computer data. (*NOAA Photo Library, NOAA Central Library; OAR/ERL/National Severe Storms Laboratory*)

- Emergency planner
- Fire weather forecaster
- Flood forecaster
- Forensic specialist
- Hurricane researcher
- Hydrological engineer
- Instrument designer
- Lightning researcher
- National laboratory researcher
- Numerical forecasting modeler
- Operational forecaster
- Paleoclimatologist
- Radar meteorologist
- Radio propagation researcher
- Remote sensing specialist
- Satellite meteorologist
- Severe storm forecaster
- Storm chaser
- Weather consultant

How do I know if **meteorology is for me**?

Choosing a career to pursue is one of the toughest choices anyone has to make. Unlike generations past, though, most Americans do not remain in their first career choice forever, and many change careers three or more times over their working lives. This is one reason why meteorology is actually a good choice, because it can be applied to a wide variety of specialties. There are some questions you can ask yourself before you decide to pursue meteorology:

1. Do I have an interest and aptitude in math, physics, and chemistry?
2. Do I have a genuine fascination for the weather and the atmosphere; do I like to read books on the subject and do so in my spare time?
3. Am I capable of conceptualizing phenomena in three-dimensional space?
4. Do I enjoy working with computers?
5. Am I flexible and willing to work in different locations and, sometimes, under a lot of pressure, such as when there is a significant storm?

You don't have to answer "Yes!" to all of these questions to become a good meteorologist. For example, number five in the list mostly applies to people in broadcasting and forecasting fields. But you should be interested in at least the majority of these and also believe you could be at least competent in the others before you seriously pursue meteorology.

Where can I **get an education** in meteorology?

There are numerous colleges and universities throughout the United States and the world where you can earn a degree in meteorology. The American Meteorological Society publishes information on this in its "Curricula in the Atmospheric, Oceanic, Hydrological and Related Sciences," which can be accessed online at http://www.ametsoc.org. The U.S. military also provides training to interested recruits. Online and correspondence courses may be found, too, but these should be viewed more for informational purposes and not for a formal education, unless the online courses are provided by a degree-conferring, accredited university.

What kinds of **courses should I take** at the university level?

For a more generalized career in meteorology, such as weather forecaster, a broad course in meteorology at the undergraduate level will prepare you well to continue on to graduate school. If you are particularly interested in detailed work in atmosphere science, you would be wise to earn an undergraduate degree in chemistry, physics, mathematics, or engineering. The growing field of environmental and climate research offers opportunities to meteorologists who also have an educational background in fields such as biology, ecology, oceanography, and geophysics. To increase your chances of finding a job at a professional level in meteorology, you will need at least a master's degree, and a doctorate is often preferred.

What was the first university to offer a degree in meteorology in the United States?

The Massachusetts Institute of Technology was the first institute of higher education to offer a meteorology degree. Swedish meteorologist Carl-Gustaf Rossby (1898–1957) founded the program there in 1928. Rossby would go on to found programs at the University of California at Los Angeles and the University of Chicago, as well.

What is a **Certified Consulting Meteorologist** (CCM)?

The American Meteorological Society (AMS) grants the title of CCM to professional meteorologists who have both broad experience in meteorology and a highly specialized knowledge of particular fields on which they intend to consult. In addition to education and experience, a meteorologist has to demonstrate professional conduct and service to be named a CCM. They are sought out by law firms, governments, law enforcement agencies, insurance companies, and other private businesses for their expertise and for the assurance that all CCMs will provide reliable and authoritative information. Only about five percent of AMS members are CCMs. The AMS provides a listing of CCMs by specialty on its website at http://www.ametsoc.org.

What is a **state climatologist**?

A state climatologist is a professional in the field who has been designated by a state government or agency of the state as the state's official climatologist. They must also be recognized by the National Oceanic and Atmospheric Administration and by the director of the National Climatic Data Center. Currently, 48 states have a state climatologist, the exceptions being Tennessee, Rhode Island, as well as Washington, D.C.; also climatologists have been designated for Puerto Rico, Guam, and the U.S. Virgin Islands. State climatologists are supported by the American Association of State Climatologists, which was founded in 1976. State climatologists receive their salaries either from the state or from a university, and they also work closely with the National Weather Service.

How do I find **financial aid** to study meteorology?

Other than contacting your university's meteorology department and its financial aid office, a good place to look for scholarships, fellowships, and internships is the American Meteorological Society. The National Council of Industrial Meteorologists also provides stipends to undergraduates. Finally, the U.S. Navy and Air Force both offer programs in meteorology for those who attend the Reserve Officers' Training Corps (ROTC) in college, and then go to Officer Training School. Your education is then paid for by the military.

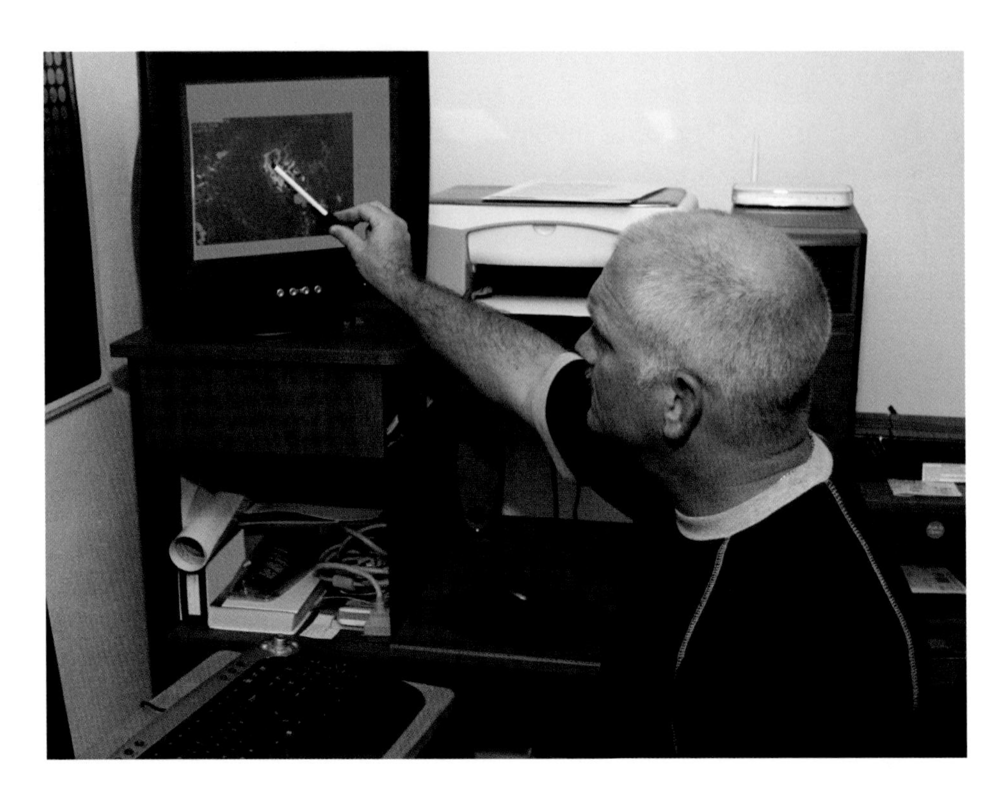

Careers in meteorology can involve a wide range of disciplines and interests, ranging from chemistry to computer science to on-air broadcasting.

How much do **meteorologists earn**?

You should never pick a career solely based on its earning potential. Pick something you love, and then go from there. That being said, meteorologists make very good wages indeed. As of 2009, average salaries ranged from $70,000 to $108,000 annually. On the low end, the salary for a degree-holding meteorologist was about $53,000; at the upper end, a meteorologist in a specialized field with a Ph.D. can easily make $125,000 a year.

Who **hires meteorologists**?

Meteorologists can either work for the government or in the private sector. If they work for the government, they may be hired by a federal government agency or department, such as the National Weather Service, the Federal Aviation Administration, the National Oceanic and Atmospheric Administration, the Environmental Protection Agency, the National Aeronautics and Space Administration, the Department of Energy, a national laboratory, or the military; or, they could be hired by state, county, or city governments. Local governments hire meteorologists to monitor such things as air pollution and other environmental and resource management concerns.

In the private sector, meteorologists can find employment at television stations (including the popular The Weather Channel), airlines, universities, utility compa-

> **What is forensic meteorology?**
>
> With the popularity of the *CSI* television series there has been a meteoric rise in students enrolling in forensic science courses. Forensics draws on a wide variety of sciences in order to solve criminal cases. For instance, there was one episode of *CSI* in which a character used his knowledge of astronomy to locate a murder scene. Meteorological principles can also be used in forensics, which is frequently applied to criminal or insurance investigations. Meteorologists have been called to testify in court, serve as consultants, or perform research for government agencies, law firms, and private businesses. Using data from satellites, radar, and other sources, a meteorologist could, for example, testify as to the possibility of a building fire being caused by lightning, or whether or not wind conditions could be responsible for an airplane crashing shortly after takeoff, or whether hazy conditions leading to a car accident were the result of nature or a nearby factory. Qualification as a Certified Consulting Meteorologist is typically required to work as a forensic meteorologist.

nies, climate research laboratories, meteorological equipment manufacturers, private research contractors and forecasting services, weather modification companies, private environmental organizations and companies, and even litigation support companies.

A good place to start a job search is the Job Board postings on the American Meteorological Society's website at http://careercenter.ametsoc.org. If at all possible, start building career connections as early as you can—preferably while you are still in school, through work internships and contacts through your professors or other people you meet.

How hard is it to become a meteorologist who is a **radio or television broadcaster**?

If your goal is to specifically become a television weatherman or weatherwoman, then, ideally, you need to combine training in meteorology with education in communications and/or the broadcast arts. You should be comfortable appearing on camera, and it helps if you have completed an internship in broadcasting while in school. As with anyone who wishes to pursue a television or radio career, you will have to sell yourself as an appealing on-air personality. Prepare professional demo tapes of yourself doing a broadcast and send these to news directors at the stations for which you wish to work. You should prepare yourself to start at the bottom of the career ladder, which means working shifts in the very early morning or very late evening hours, as well as working for low pay in a town or city that has a small market. Don't expect to work right away for a big network in New York City or Los Angeles. It could be years before you manage to get a job in a big market, and in the meantime you will likely move frequently while you search for better and better opportunities. Today's media world is especially challenging, as stations and newspapers across the United States are cutting their budgets and centralizing operations. Weather broadcasting is probably the hardest specialty to get into these days, and many other fields within meteorology offer better opportunities.

Why do **environmental firms** hire meteorologists?

Weather patterns have an important effect on the distribution of pollutants in the air, on land, and in our oceans. Environmental firms, as well as government agencies such as the Environmental Protection Agency, hire meteorologists to help them predict the environmental impact of construction projects such as power plants and factories. An understanding of prevailing winds near a proposed coal-burning plant, for instance, will help people understand how potential air pollution, acid rain, and ozone levels will impact the environment not only locally but also, perhaps, across many states or even countries.

To what **professional societies and organizations** should a meteorologist belong?

Most meteorologists in the United States participate as active members of the American Meteorological Society. Those specifically interested in the atmospheric sciences often also join the American Geophysical Union, which is headquartered in Washington, D.C. Meteorologists specializing in forecasting can join the National Weather Association for its benefits; the NWA is also concerned with the operational aspects of meteorology.

Index

Note: (ill.) indicates photos and illustrations.

B

D

E

F

G

H

I

J

K

L

N

O

P

R

S

T

U

X, Y, Z